FUNGICIDES FOR FIELD CROPS

FUNGICIDES FOR FIELD CROPS

Frank Spacey
Christopher Holland
Sophia Dunkley
Issy Wong

Kruger Brentt
Publishers
2023

Kruger Brentt Publishers UK. LTD.
Company Number 9728962

Regd. Office: 68 St Margarets Road, Edgware, Middlesex HA8 9UU

ISBN: 9781787150584

For information on all our publications visit our website at http://krugerbrentt.com/

PREFACE

Fungi can be difficult to control. They produce a massive quantity of spores which can spread via rain or wind over long distances to settle in favourable conditions such as surfaces of leaves, stems or fruit. Fungi can grow through and between plant cells, withdraw nutrients, and even in some cases produce toxins that kill the plants. Fungi can also thrive in crop debris and infect the next cycle of crops. Fungicides for plants can prevent the fungi spores from growing and eliminate the fungi that might already exist to protect the health of the crops. Modern fungicides have been developed to specifically target fungi and do a lot more. Synthetic chemical fungicides can provide protection without being absorbed or translocated by the plant. Fungicides also make it possible to provide protection against a wide variety of fungi. Some systemic fungicides have also been developed that can be absorbed by the plant and have a therapeutic effect by curing plant diseases. Apart from synthetic fungicides, biofungicides can also be used to control certain types of fungi and are especially useful for organic production. Fungicides are a necessary part of today's agricultural sector as without its production would be severely impacted as there would be no protection from diseases.

Fungicides are pesticides that kill or prevent the growth of fungi and their spores. They can be used to control fungi that damage plants, including rusts, mildews and blights. They might also be used to control mold and mildew in other settings. Fungicides work in a variety of ways, but most of them damage fungal cell membranes or interfere with energy production within fungal cells. Plant pathogenic fungi cause devastating damage to crop production worldwide. The growing global population necessitates reduced crop losses to improve food security, and the control of fungal plant pathogens is vital to help maintain food production.

The present book contains eight chapters covering all related disciplines. These chapters include *Introduction, plant pathology and plant pathogens, botanical fungicides in agriculture, decision-making factors for fungicide use, role of fungicides in agriculture, and horticulture, fungicide stewardship, use of copper-based fungicides in organic field crops, using fungicides to manage diseases on field crops.*

This book highlights the importance of fungicides for yield protection, while considering economics, risks to humans and the environment, and prolonging use of fungicide modes of action. This book will serve as a valuable resource for agribusiness professionals, researchers and extension personnel, farmers and crop production advisors, and teachers and students.

We are grateful to all those persons as well as various books, manuals, periodicals, magazines, journals etc. that helped in the preparation of this book. In spite of the best efforts, it is possible that some errors may have occurred into the compilation and editing of the book. Further queries, constructive suggestions and criticisms for the improvement of the book are always welcome and shall be thankfully acknowledged.

Frank Spacey
Christopher Holland
Sophia Dunkley
Issy Wong

CONTENTS

1

INTRODUCTION

Fungicides are pesticides that kill or prevent the growth of fungi and their spores. They can be used to control fungi that damage plants, including rusts, mildews and blights. They might also be used to control mold and mildew in other settings. Fungicides work in a variety of ways, but most of them damage fungal cell membranes or interfere with energy production within fungal cells. Fungicides, herbicides and insecticides are all pesticides used in plant protection.

A fungicide is a specific type of pesticide that controls fungal disease by specifically inhibiting or killing the fungus causing the disease. Not all diseases caused by fungi can be adequately controlled by fungicides. These include the vascular diseases Fusarium and Verticillium wilt. Diseases caused by other types of organisms, disorders caused by abiotic factors, and insect damage are not controlled by fungicides. Thus it is essential to first determine the cause of symptoms before applying a fungicide.

Diseases are a common occurrence on plants, often having a significant economic impact on yield and quality, thus managing diseases is an essential component of production for most crops. Broadly, there are three main reasons fungicides are used:

a. To control a disease during the establishment and development of a crop.
b. To increase productivity of a crop and to reduce blemishes. Diseased food crops may produce less because their leaves, which are needed for photosynthesis, are affected by the disease (Fig. 1-2). Blemishes can affect the edible part of the crop or, in the case of ornamentals, their attractiveness, which both can affect the market value of the crop.
c. To improve the storage life and quality of harvested plants and produce. Some of the greatest disease losses occur post-harvest (Fig. 3).

Fungi often spoil (render unusable) stored fruits, vegetables, tubers, and seeds. A few which infect grains produce toxins (mycotoxins) capable of causing severe illness or even death in humans and animals when consumed. Fungicides have been used to reduce mycotoxin contamination in wheat affected by Fusarium head blight, but most fungicides developed so far have not been sufficiently effective to be useful for managing mycotoxins associated with other diseases.

Fig. 8. The disease early blight has killed most of the leaves on the tomato plants on the right, while nearby plants that were treated with fungicides (left) have maintained their healthy leaves and will produce many more fruit and fruit of higher quality.

Fig. 2. Apple trees sprayed with fungicides during summer have healthy foliage during early autumn (left) whereas foliage on unsprayed trees is attacked by various fungi that promote premature defoliation (center and right). Premature loss of leaves in October results in fewer fruit the following year.

Fungi and oomycetes are the causal agents of numerous fungal plant diseases, causing losses in agricultural production worldwide. Plant pathogenic fungi and oomycetes have coevolved with their plant hosts, and have developed extremely efficient mechanisms to cause an infection, to grow, multiply and spread on living plants. Most of these plant pathogens have an exceptional potential to reproduce within the plant host. For example, Pytophthora infestans can produce 50 000 to 117 000 sporangia per linear centimeter of infected potato stem length.

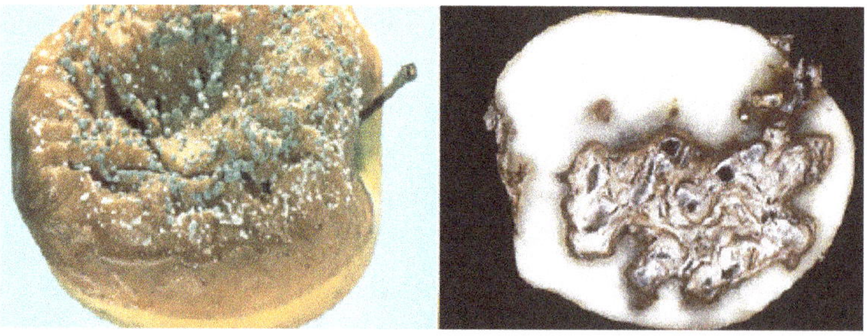

Figure 3. Post-harvest diseases that can cause extensive losses include blue mold affecting apple (left) and Fusarium dry rot of potato (right).

Larger apothecia of Sclerotinia trifoliorum can release about 4.7 million ascospores during their active period . In late vegetation stages, the number of Fulvia fulva conidia per square centimeter of infected tomato leaf can exceed 17 000, while Pseudoperonospora cubensis can produce even 195 000 sporangia per square centimeter of cucumber leaf seven days after infection. Considering the prevalence, ubiquitous nature and the ability of fungi and oomycetes to cause epidemics in relatively short period of time, disease management strategies are needed to secure the productivity in today's agriculture.

Chemical control measures are particularly common in management of fungal plant diseases, and most of these measures rely on the use of fungicides. No matter of many differences in biology of different fungal and oomycetous plant pathogens, they share certain similarities. Most of the fungi produce various kinds of spores, which come in contact with plant tissue, germinate, and penetrate into the plant during the infection process. Some plant pathogenic fungi (e.g. *Rhizoctonia solani, Sclerotium rolfsii,* or *Macrophomina phaseolina*) rarely produce spores and infect plants through penetration of mycelia into plant tissue. After infection, fungi and oomycetes continue to grow as mycelium in or on plant organs, where they produce their new vegetative or generative, propagation or dormant structures.

Nearly all fungicides used in agriculture today show their best effect if applied before the infection occurs. When present on the surface of plant organs, fungicides destroy fungal spores or suppress germination tubes, hyphae, and other fungal structures. As a rule, control of fungal diseases with fungicides is aimed to prevent an infection and subsequent disease development, and in such way the use of fungicides in plant protection differ from the use of antibiotics or antimycotics in medicine and animal health. As fungal diseases are a major threat to crop production:

1. the application of fungicides to control fungal infestations is often considered indispensable to secure global food supply.
2. In the European Union (EU), fungicide sales (based on mass) account for more than 40% of the total pesticide sales with synthetic, organic fungicides accounting for approximately 60% of all fungicides.
3. In wine-growing regions, fungicides can account for more than 90% of all pesticide applications.
4. Moreover, fungicide use is regionally predicted to increase because of changes to climatic conditions, development of fungicide resistance, and invasive fungal species.

Fungicides can enter aquatic ecosystems via point (e.g., discharge from wastewater treatment plants following domestic and agricultural use) and nonpoint (e.g., drift, drainage, surface runoff primarily from agricultural use) sources. In aquatic systems, fungicides can be toxic to a wide range of nontarget organisms as they act on basic biological processes that are not specific to fungi (e.g., energy production).

Despite intensive use of fungicides and the associated potential ecotoxicological risks in nontarget aquatic systems, the environmental fate and effects of fungicides have received far less attention compared to insecticides and herbicides. For instance, only 13% of studies on pesticide effects between 1991 and 2013 focused on fungicides, compared to 62% and 24% for insecticides and herbicides, respectively. Therefore, we provide an overview of the risk of fungicides to aquatic ecosystems.

CLASSIFICATION OF FUNGICIDES ACCORDING TO MOBILITY

Non-penetrating Fungicides—Non-Mobile

Non-penetrating fungicides are not absorbed by the plant tissues and therefore not translocated. These fungicides remain on plant surfaces as a protective barrier or "shield" that will inhibit spore germination and/or mycelial growth. Therefore, this type of fungicides can be removed from the plant surface by rain or irrigation. Some examples of non-penetrating fungicides that have been formerly used to control SR include: mancozeb, sulfur, maneb, bordeaux mixture, cupric salts, nickel salts, etc..

Penetrant and Mobile (Systemic) Fungicides

Penetrating fungicides are deposited on plant surfaces, get absorbed by leaf tissues and transported upward by xylem vessels. They may be transported at very short distances (local systemic movement) or over long distances within the plant (true systemic fungicide) from the site of uptake depending on chemical compound attributes.

Mobile fungicides have contributed to improve crop disease control since, unlike non-penetrating fungicides, are not exposed to leaching and photo-decomposition, requiring fewer number of applications. Thus, penetrant-mobile fungicides present high control efficiency in smaller rates per hectare in comparison with non-penetrating fungicides.

Among penetrant-mobile fungicides, those that are more mobile within plant tissues, are transported upward by xylem (acropetal movement). Only a few fungicides can be partially translocated basipetally in the phloem.

The term "mesostemic" was devised to identify a complex group of strobilurins fungicides with diverse characteristics. Mesostemic movement refers to penetrating fungicides that act and move in the mesophyll. Some of them are also translaminar (passes from one leaf face to the other) and systemic (ex. azoxystrobin, picoxystrobin) while others also move through the vapor phase, e.g., trifloxystrobin.

CLASSIFICATION OF FUNGICIDES ACCORDING TO MODE OF ACTION

Fungicides interfere with several cellular processes essential to the life of fungi and fungal-like organisms. How a fungicide molecule inhibits the growth or kills a given fungus is refer to its mode or mechanism of action (MOA) and it constitutes one of the way of classification. Currently, Fungicide Resistance Action Committee (FRAC) scheme lists 11 different modes of action (FRAC, 2020).

Different MOAs, applied as seed or foliar treatments, are used to prevent and cure SR. Non-penetrating protective fungicides have been used several years ago in SR control . Currently, systemic fungicides such us DMI (FRAC Code 3), QoI (FRAC Code 11), and succinate dehydrogenase inhibitors (SDHI) (FRAC Code 7) are the most widely used fungicides worldwide.

According to their chemical structure, within the DMI fungicides the most numerous and important group are the triazoles. Although the fungicides strobilurins and carboxamides have the same MOA (inhibition of mitochondrial respiration), they have different sites of action. Carboxamides act in complex II whereas strobilurins act in complex III.

On the other hand, triazoles inhibit ergosterol biosynthesis, an essential component of the fungal cell membrane. Strobilurins, carboxamides and triazoles are single-site fungicides, i.e., they act against a single target or essential enzyme in an important metabolic pathway in a fungus. In contrast, fungicides that inhibit multiple sites, such as maneb, mancozeb, and chlorothalonil, affect several metabolic sites within a fungus.

The most frequent used fungicides to control SR are DMIs: cyproconazole, epoxiconazole, fluquinconazole, flutriafol, metconazole, propiconazole, prothioconazole, and tebuconazole; QoIs: azoxystrobin, kresoxim methyl, trifloxystrobin picoxystrobin, and pyraclostrobin; and SDHIs: fluxapyroxad, bixafen, and benzovindiflupyr. Particularly, DMI fungicides have been used extensively to control SR.

CLASSIFICATION OF FUNGICIDES ACCORDING TO SUB-PHASES OF THE AFFECTED INFECTIOUS PROCESS

In this classification scheme, fungicides are grouped according to which subphase or event of the pathogenesis is affected. Pathogenesis or disease cycle is a series of dynamic events that occur in succession during a parasitic relationship of a pathogen and a host that leads to development and establishment of infection. The complete disease cycle includes spore dispersal and deposition, spore germination and penetration, infection, host colonization and invasion, reproduction, dissemination, and survival. The incubation period is the time from the beginning of infection till the appearance of first symptoms. Latent period is the time from the beginning of infection until the appearance of first signs. Based on which sub-phase of the infectious process is affected, fungicides can be classified as preventive, curative, and eradicant.

PROTECTANT OR PREVENTIVE FUNGICIDE

Preventive or protectant fungicides act before fungal spores are deposited or before spore germination occurs. The main action exerted by the fungicide is "protectant" or "pre-penetration." The fungicide prevents penetration and infection. All non-penetrating fungicides should be considered preventive o protectant agents.

Some penetrating fungicides (strobilurins and carboxamides) can also have preventive or protective action . This is explained by their mechanism of action based on the inhibition of mitochondrial respiration, a process that is critical during germination of spores.

Curative—Penetrating Fungicide

The fungicide is able to inhibit fungal growth inside the plant tissues before symptoms and signs are observed. They act mainly during the incubation period, paralyzing the infectious process. Disease control occurs after infection but without the presence of symptoms. Triazoles are typically curative fungicides and are frequently used in the control of rusts.

Eradicant Fungicides—Penetrating Fungicide

The eradication activity of a fungicide is related to the inhibition of disease progress after the appearance of symptoms or signs. Complete eradication of the pathogenic fungus within the host tissues is rare and difficult to achieve in the field. Most fungicides do not have a significant eradication action, being their preventive and/or curative activities the main attributes to achieve efficient controls.

FUNGICIDE SEED TREATMENTS

Systemic fungicides applied on seeds are up-taken by seed tissues and seminal roots during seed germination and seedling establishment and then mobilized through the xylem to the plumule and seedling leaves. Thus, certain systemic fungicides are used as seed treatments to protect seedling leaves exposed to early deposition of SR spores.

Fungicide seed treatments for SR control is beneficial especially in regions where highly susceptible varieties are grown, the disease is frequent, and *Pst* attacks occurs at early wheat vegetative stages. Sometimes, fields that were planted to seed efficiently treated against SR can delay or decrease the number of foliar applications. The major problem with fungicide seed treatment is the plausible phytotoxic effects on the plant stand, especially when high doses of a.i. are used.

The molecules evaluated were oxathiin penetrant-mobile fungicides: carboxin, oxicarboxin and several substituted analogs of carboxin. Subsequently, other molecules, such as butrizol, triadimenol, triadimefon, were evaluated. Triadimefon has shown high efficacy in controlling SR and has been one of the most used fungicides in China. Another DMI fungicide, such as fenpropimorph (morpholine, FRAC Code 5) was also effective for SR control.

Currently, some triazoles such us triticonazole, flutriafol, fluquinconazole and the new carboxamide fluxapiroxad that showed high control efficiency are widely recommended as a seed treatment for SR control. Seed treatment for SR control should be considered as part of the integrated disease program.

FOLIAR FUNGICIDE APPLICATION

The first fungicides that provided effective SR control were non-penetrating and protective. However, their use was limited due to the need to be applied repeatedly and because of their ineffectiveness on established infections. Other fungicides, such as nickel salts and even antibiotics, such as phleomycin gave satisfactory SR control. A great advance in the development of fungicides for the control of SR was the discovery of oxathiin fungicides, especially oxycarboxin, which significantly improved SR control. However, in field trials conducted in Canada, oxycarboxin was not effective for the control of SR.

Subsequently, the development of new systemic molecules increased the possibilities of chemical control. In the USA, since the 1970s numerous tests have been carried out to assess different fungicide molecules and their control efficacy against SR. A milestone in the history of SR chemical control was the devastating epidemics of both leaf and stripe rust that occurred in the U.S. Pacific Northwest in 1980 and 1981. The epidemic forced the registration of the fungicide triadimefon, which had proved to be the most efficient molecule in annual field trials. Use of this fungicide reduced and avoided large economic losses due to SR.

Similarly, faced with the emergence of unexpected SR epidemics caused by new exotic *Pst* races, it also took time to register fungicides for SR control in several countries, such as Canada. Between 1981 and 1986, triadimefon was made available in Alberta to control SR but on a temporary basis. Afterwards, propiconazole was given restricted registration for SR control.

The search for new and efficient fungicidal molecules and the assessment of their greenhouse and field efficacies continued incessantly. Since the 1980s, DMIs began to be protagonists not only for the control of SR but also for other foliar fungal diseases. Strobilurins were introduced at the end of the 1990s and were widely disseminated in most crops. Although numerous cases of fungicide resistance to this chemical family have emerged (FRAC, 2020), they still maintain efficiency against rusts.

One exception is the Asian soybean rust, for which sensitivity reduction has been reported. Since ~2010, the SDHIs have been introduced in all fungicide markets in the world. This fungicide group successfully complemented QoIs and DMIs to control the complex of fungal diseases that infect wheat. Nevertheless, they are classified by FRAC as high risk fungicides.

Currently, there is a great diversity of fungicide commercial formulations containing a single or several active ingredients in mixture that are registered and/or recommended for SR control in all wheat-growing regions in the world. Although the fungicide active ingredients may differ in their effectiveness according to the field dose, most of the registered fungicides have shown high SR control when properly applied. The QoI + DMI mixture or mixtures with SDHIs are excellent options because in addition to controlling SR they extend control to other wheat diseases.

For example, in Argentina, the use of fungicides for SR control was based on DMIs and their mixtures with QoIs. The DMIs alone showed high SR control, but their persistence period was shorter than when used in mixtures. In addition, they did not provide preventive action. When SR appeared in together with leaf rust (*Puccinia triticina*), DMIs were not efficient in controlling it because *P. triticina* has decreased sensitivity to this group of fungicides. Likewise, field observations showed that DMIs can stop new SR infections, but cannot efficiently control infections older than 1 week or more, which will develop into necrotic stretch marks visually appearing as "a great necrotic area". Therefore, mixtures of fungicide active ingredients proved to be a better option to SR control by increasing the period of protection, granting preventive and curative action while also being able to control both rusts with high efficiency

ROLE OF FUNGICIDES IN DISEASE MANAGEMENT

Plant diseases are best managed by integrating a number of control practices that may include: crop rotation, selection of disease-tolerant or disease-resistant crop cultivars (cultivars genetically less susceptible than other cultivars), time of planting, level of fertilization, micro-climate modification, sanitation, and application of fungicides. Fungicides are often a vital part of disease management as:

a. They control many diseases satisfactorily,
b. Cultural practices often do not provide adequate disease control,
c. Resistant cultivars are not available or not accepted in the marketplace for many diseases, and
d. Certain high value crops have an extremely low tolerance for disease symptoms.

in contrast with most human medicines, most fungicides need to be applied before disease occurs or at the first appearance of symptoms to be effective. Unlike with many diseases of humans and animals, applying fungicides cannot heal symptoms already present, even if the pathogen is killed. This is because plants grow and develop differently than animals. Fungicides typically only protect new uninfected growth from disease.

Few fungicides are effective against pathogens after they have infected a plant. Those that do have "curative" properties, which means they are active against pathogens that have already infected the plant, have limited ability to do so, often only being active on a pathogen within a few days of infection.

Many fungicides have targeted activity that imparts high efficacy against specific pathogens, which means low potential for toxicity to humans and other organisms, but also results in a high risk of pathogens developing resistance to the fungicide. A resistant pathogen is less sensitive to the action of the fungicide, which results in the fungicide being less effective or even ineffective.

Fungicides that are designed to target specific enzymes or proteins made by fungi do not damage plant tissue, thus they can penetrate and move inside leaves enabling curative properties and increasing the amount of plant tissue protected to more than just where fungicide was deposit when applied.

Since the mode of action of these fungicides is so specific, small genetic changes in fungi can overcome the effectiveness of these fungicides and pathogen populations can become resistant to future applications. Disease management strategies that rely heavily upon curative application of fungicides often lead to more resistance problems due to

a. The large size of the pathogen population when the application is made from which resistant individuals are being selected and
b. The difficultly in eradicating a pathogen entirely from inside the plant. Fungicide resistance is covered in more detail in a separate section.

Growers often use disease forecasting systems or action thresholds, when these are available, to ensure fungicides are applied when needed and to avoid the expense and possible environmental impact of unnecessary applications. Forecasting systems have been developed for a number of diseases based on an understanding of the environmental conditions favorable for their development. Typically these are based on temperature and relative humidity or leaf wetness in the area where the crop is grown.

Threshold-based fungicide programs involve routinely scouting the crop for symptoms, then applying fungicides when the amount of symptoms reaches a critical level beyond which the disease cannot be controlled adequately. An example of a critical level is one disease spot per five leaves examined. Knowledge of the disease cycle of the pathogen is important when developing and using forecasting systems and thresholds. Important aspects of the disease cycle include whether the disease is monocyclic (one generation per year) or polycyclic (multiple generations) and latent period (time between infection and production of new inoculum).

Economics often influence the choice of fungicide and application timing. Expensive fungicides and numerous applications are used on valuable plantings that might incur substantial economic loss in the absence of treatment, such as fruit trees and golf courses. Recognizing that with some diseases crop yield is not impacted when severity is low, an economic threshold is used to determine when fungicide treatment is needed.

The crop tolerance level, or damage threshold, can vary depending upon the stage of the crop development when attacked, crop management practices, location and climatic conditions.

APPLICATION METHODS

Multiple organisms (viruses, nematodes, fungi, and bacteria) can cause plant disease. Preventing and managing disease is best accomplished by a combination of practices, known as Integrated pest management or IPM. Management practices include matching the plant with the site, selecting disease-resistant varieties, plant care that prevents stress (irrigation, mulch, fertilization as needed etc.) as well as fungicide use when warranted.

Fungicides are pesticides that prevent, kill, mitigate or inhibit the growth of fungi on plants, but they are not effective against bacteria, nematodes, or viral diseases. Fungicides can be classified based on:

- **Mobility in the plant:** Contact vs. mobile (types of systemics). Contact fungicides (AKA protectants) are not absorbed by the plant and stick to plant surfaces. They provide a protective barrier that prevents the fungus from entering and damaging plant tissues. Systemic products (also known as penetrants), are absorbed by the plant and can move from the site of application to other parts of the plant. Movement in the plant varies by fungicide, form moving to old and new tissues (amphymobile or true systemic), new growth (acropetally or xylem mobile), moving from the top to the bottom of the leaf surface (translaminar). For more information see the publication "Fungicide Mobility for Nursery, Greenhouse, and Landscape Professionals."
- **Preventive vs. curative:** Preventive fungicides work by preventing the fungus from getting into the plant. The preventive fungicide must come into direct contact with the fungus, and they have to be re-applied to new plant tissues (as leaves or needles expand in the spring) or if the product washes off. Curative fungicides affect the fungus after infection. This means they can stop the disease after the infection has started or after first symptoms are observed. Fungicides that can move in the plant can be both preventative and curative.
- **Mode of action:** This refers to how the fungicide affects the fungus. Fungicides may work by damaging the cell membrane of the fungus, inhibiting an important process that the fungi, pinpointing a single or multiple processes in the fungus. It's important to incorporate different modes of action by mixture or by alternating products to maintain effectiveness and prevent fungicide resistance. Stay tuned for our article "what is fungicide resistance?".

Fungicide use appears to be the fastest growing segment. Here is some advice on how to get the best bang for the buck.

- Timing is the most important part of fungicide application. Diseases can develop and spread quickly. Most fungicides cannot cure a disease infection, they can only protect against it. If an application misses the window, yield is lost. Remember your priorities – become familiar with disease symptoms, the susceptibility of your crop and key growth stages. Make sure your sprayer is ready – your nozzles are installed, calibrated, and you can achieve the necessary boom height. Hire an agronomist to help scout and make recommendations. Make the right decision about whether to spray or not.

- Water volume is the most important application parameter for fungicide application. In years of study, increasing water volume had a greater effect on fungicide performance than changes in droplet size or spray pressure. More water is needed for fungicides than herbicides because of the greater amount of plant material present. Getting coverage on leaf areas deeper into the canopy requires more water. Although finer sprays can also help with coverage, this practice is riskier due to drift potential and higher evaporation rates.
- Double nozzles, in particular the asymmetric types, are becoming more popular with fungicides. Double nozzles are proven effective and recommended primarily for fusarium head blight, or any other disease where an exposed vertical part of the plant canopy is the primary spray target. Double nozzles are also useful for preventing the spray quality from getting too coarse as higher flow-rate nozzles (which tend to have larger droplets) are used.
- Travel speed is important with fungicides. Canopy penetration sometimes improves with slower travel speeds, and this can be used as an advantage by eliminating the need for a special fungicide nozzle. For example, assume a nozzle was used to apply 8 gpa of herbicide at 15 mph at 70 psi (this pressure assumes air-induced tips). For fungicides, this same nozzle and pressure will deliver 12 gpa simply by slowing down to 10 mph.

Fig. 4 Double nozzles

Boom height and spray quality are both important for single angled sprays or double nozzles. The angle at which a spray leaves a nozzle diminishes quickly as air resistance and gravity exert their influence. If the boom is too high, the initial forward angle will be lost and the spray droplets will actually deposit with gravity and wind. But if the spray is a bit coarser and the boom is low enough, the angle of attack is retained for long enough to make a difference in spray deposition.

Despite these suggestions for making the spray more effective, there is no substitute for an informed decision regarding fungicide use. It's possible that spraying is unnecessary for a number of reasons, and it's best to have professional advice help make that call. If you decide to go ahead, ensure that your sprayer is set up to deliver the fungicide to the part of the canopy that needs protection.

RULES OF THUMB FOR FUNGICIDE USE

The problem has to be diagnosed correctly: Before applying a fungicide make sure that you know the cause of the disease (is it a disease? If so what is causing it? Fungi? which one?) and when (spring, fall etc.) and how often to apply the fungicide. The timing of the fungicide application can enhance the effectiveness of the product and prevent additional sprays.

When ready to use the recommend fungicide for the particular problem your plant is facing, read the label and follow instructions. This will not only protect your plant, but it will also protect your health and the environment. Remember always to apply fungicides using the appropriate equipment at the recommended application rate.

Fungicides are applied as dust, granules, gas, and, most commonly, liquid. They are applied to:

a. Seed, bulbs, roots of transplants, and other propagative organs. These treatments are usually done by the seed company. Some treatments need to be done by the grower on-site at the time of planting. The goal is to kill pathogens that are on the planting material or to protect the young plant from pathogens in the soil.
b. Soil either in-furrow at planting, after planting as a soil drench (including through drip irrigation), or as a directed spray around the base of the plant.
c. Foliage and other aboveground parts of plants by means of a sprayer.
d. Inside of trees via trunk injection.(f) Harvested produce, as a dip or spray in the packinghouse.

Fungicides are used as a formulated product consisting of an active ingredient plus inert ingredients that improve the performance of the product. Fungicides are typically mixed with water then applied by spraying. Application equipment ranges from small hand-held and back-pack sprayers to large spray units carried by tractors or aircraft. A few fungicides are applied as dusts. Fungicides can also be applied in greenhouses as smoke, mist, fog or aerosol. Coverage of all parts of the plant susceptible to the disease is critical because very few fungicides can move adequately throughout a plant. Advancements are continually being made to nozzles and sprayers to improve coverage.

For many diseases, effective control necessitates multiple applications of fungicides, sometimes as frequently as every 5 days. Repeated applications are needed to protect new growth and to replace fungicide lost from the plant by chemical decomposition, UV-light degradation, and erosion by wind and water.

TYPES OF FUNGICIDES AND DEFINITIONS

Fungicides are categorized in several ways based on different characteristics.

1. **Mobility in a plant:** contacts or systemics: Contact fungicides (also called protectants) remain on the surface of plants. Many contacts are potentially phytotoxic (toxic to plants) and can damage the plant if absorbed. Systemics (also called penetrants and mobile fungicides) are absorbed into plants. Most systemics move very short distances from the site of application, such as across a leaf blade from one surface to the other (local systemic or translaminar). Some fungicides are weakly systemic and can move further from the application site than local systemics, reaching all parts of the leaf the fungicide is deposited on. A few systemics move more extensively because they are mobile in xylem tissue. When applied to the root zone, these are absorbed by roots and then move upward through the plant with the transpiration stream (xylem-mobile systemic). Xylem-mobile systemics applied to leaves move throughout the leaf where deposited, but cannot be redistributed out of that leaf; however, any material deposited on stems can move upwards into leaves. Phloem-mobile systemics (also known as "true" or amphimobile systemics) have bi-directional mobility, some material moves in phloem out of the leaf where deposited upwards to other leaves and downwards to roots. Systemics cannot move again after translocation.
2. Role in protection: **preventive or curative. Contacts are products suited for** preventive (prophylactic) use as they work by contact action on the surface of the plant to which they have been applied. Repeated applications are needed to protect new growth of the plant and to replace material that has been washed off by rain or irrigation, or degraded by environmental factors such as sunlight. Sometimes contacts are referred to as "residual" products because the deposited fungicide remains on the plant surface, occasionally as a visible residue, for several days. Due to their ability to penetrate plants, some systemics possess both preventive and curative (eradicant or kick-back) activity, thus affecting the pathogen after infection.
3. **Breadth of activity:** single-site or multi-site. Single-site fungicides are active against only one point in one metabolic pathway in a pathogen or against a single critical enzyme or protein needed by the fungus. Since single-site fungicides are highly specific in their toxicity, having little effect on most organisms, they can safely be absorbed into plants, thus these fungicides tend to have systemic properties. As a result of this specific activity, fungi are more likely to become resistant to the fungicide because a single mutation in the pathogen usually allows it to overcome the action of the fungicide, such as by preventing it from binding to the active site in the fungus.

- Typically, older contact fungicides have multi-site activity and thus usually affect many fungi in different classes. Through the development of in vivo screens, and due to the increase in the stringency and number of regulatory tests required to register a new active ingredient, fungicide manufacturers have found it easier to develop single-site systemics recently. As a result, fungicide resistance has become a more important concern in disease management.

Mode of Action

Fungicides kill fungi by damaging their cell membranes, inactivating critical enzymes or proteins, or by interfering with key processes such as energy production or respiration. Others impact specific metabolic pathways such as the production of sterols or chitin. For example, phenylamide fungicides bind to and inhibit the function of RNA polymerase in oomycetes, while the benzimidazole fungicides inhibit the formation of beta tubulin polymers used by cells during nuclear division.

Some recently developed products are unique in that they do not directly affect the pathogen itself. Many of these elicit a response from the host plant known as "systemic acquired resistance" (SAR). These SAR inducers basically mimic chemical signals in plants that activate plant defense mechanisms such as the production of thicker cell walls and anti-fungal proteins. The utility of SAR inducers, however, has been limited so far since many pathogens are capable of over-powering such defenses.

Knowledge of exactly how a fungicide affects a fungus is helpful for selecting products. First, mode of action determines which fungi will be affected by a fungicide and thus which diseases can be controlled by using the fungicide. Second, fungicides with different modes of action are needed in a disease management program to delay fungicide resistance development.

Breadth of activity: narrow-spectrum or broad-spectrum. Narrow-spectrum fungicides are effective against only a few usually closely related pathogens. These usually have single-site activity and are often systemic. Broad-spectrum fungicides can often control a wide range of unrelated pathogens. These usually are contacts with multi-site activity, but some have single-site activity. Several fungicides have activity that places them on a continuum between narrow- and broad-spectrum fungicides.

Type of chemical: inorganic or organic. Fungicides also can be classified based upon their chemical composition. Chemically, organic molecules are those that contain carbon atoms in their structure whereas inorganic molecules do not. Many of the first fungicides developed were inorganic compounds based on sulfur or metal ions such as copper, tin, cadmium and mercury that are toxic to fungi. Copper and sulfur are still widely used. Most other fungicides used today are organic compounds and thus contain carbon. The term "organic" as used here is based on chemistry terminology and differs from "organic" used to describe a system of agriculture that strives to be holistic and to enhance agroecosystem health.

ADDITIONAL TERMS

- Active ingredient (a.i.) is the active component of a pesticide, and is that moiety which is patented, synthesized, and registered by the manufacturer.
- Biofungicide is a naturally based microbial or biochemical product. There are three types of biopesticides. (a) Microbial biopesticides have an active ingredient that is a biological control agent (organism capable of attacking or competing with a pathogen or pest). (b) Plant biopesticides or plant-incorporated protectants are defined by the EPA as "pesticidal substances that plants produce from genetic material that has been added to the plant". (c) Biochemical biopesticides contain naturally-occurring substances. Some biochemicals may also be products of fermentation. Biochemicals can directly affect disease-causing organisms or may stimulate SAR. Biopesticides generally are narrow-spectrum, have low toxicity, decompose quickly, and thus are considered to have low potential for negative impact on the environment. Many biofungicide products are being approved for organic crop production (www.omri.org). While many have low toxicity, biopesticides are not necessarily safer than pesticides containing synthetic ingredients.
- Chemical group or class is the name given to a group of chemicals that share a common biochemical mode of action and may or may not have similar chemical structure.
- Chemical name is denoted by a nomenclature system designed by the International Union of Pure and Applied Chemistry (IUPAC); this set of rules is used to name organic, carbon-based compounds.
- Common name is the name proposed by the basic manufacturer for the active ingredient. This name is ratified by a committee either at IUPAC or the ACS (American Chemical Society).
- Reduced risk is a special classification approved by the EPA for specific uses of pesticides that have low risk to human health, low toxicity to non-target organisms, low potential to contaminate water or other environmental resources, and/or that broaden the adoption and effectiveness of integrated pest management strategies. Registration is expedited for uses designated as reduced risk.
- Formulation is the pesticide product. It is a prepared mix of active ingredient plus ingredients that improve pesticidal performance, such as carriers, solvents, wetting agents, spreaders, and stickers. Types of formulations include wettable powders, dry flowables, emulsifiable concentrates, and granules.
- Trade name/Trademark is the patented name under which a product is sold to the end user.

FUNGICIDE RESISTANCE

Fungicide resistance is a stable, heritable trait that results in a reduction in sensitivity to a fungicide by an individual fungus. This ability is obtained through evolutionary processes. Fungicides with single-site mode of action are at relatively high risk for resistance development compared to those with multi-side mode of action. Most fungicides being developed today have a single-site mode of action because this is associated with lower potential for negative impact on the environment, including non-target organisms.

When fungicide resistance results from modification of a single major gene, pathogen subpopulations are either sensitive or highly resistant to the pesticide. Resistance in this case is seen as complete loss of disease control that cannot be regained by using higher rates or more frequent fungicide applications. This type of resistance is commonly referred to as "qualitative resistance".

When fungicide resistance results from modification of several interacting genes, pathogen isolates exhibit a range in sensitivity to the fungicide depending on the number of gene changes.

Variation in sensitivity within the population is continuous. Resistance in this case is seen as an erosion of disease control that can be regained by using higher rates or more frequent applications. Long-term selection for resistance in the pathogen by repeated applications may eventually result in the highest labeled rates and/or shortest application intervals not being able to adequately control the disease. This type of fungicide resistance is commonly referred to as "quantitative resistance".

Fungal isolates that are resistant to one fungicide are often also resistant to other closely-related fungicides, even when they have not been exposed to these other fungicides, because these fungicides all have similar mode of action. This is called cross resistance.

Fungicides with the same Group Code are likely to exhibit cross resistance. Occasionally negative cross resistance occurs between unrelated fungicides because the genetic change that confers resistance to one fungicide makes the resistant isolate more sensitive to another fungicide.

Managing fungicide resistance is critically important to extend the period of time that an at-risk fungicide is effective. The primary goal of resistance management is to delay its development rather than to manage resistant fungal strains after they have been selected. Therefore, resistance management programs need to be implemented when at-risk fungicides first become available for commercial use. The objective of resistance management is to minimize use of the at-risk fungicide without sacrificing disease control. This is accomplished by using the at-risk fungicide with other fungicides and with non-chemical control measures, such as disease resistant cultivars, in an integrated disease management program.

It is critical to use an effective disease management program to delay the build-up of resistant strains. At-risk fungicides should be used at the manufacturer's recommended rate (full rate) and application interval. Using full rates is expected to minimize selection of strains with intermediate fungicide sensitivity when resistance involves several genes (quantitative resistance). At-risk fungicides should be used in alternation with other at-risk fungicides with different modes of action or different chemical groups, and they should be combined or alternated with fungicides that have a low resistance risk.

When one crop could serve as a source of inoculum for a subsequent crop, the alternation scheme among at-risk fungicides should be continued between successive crops such that the first at-risk fungicide applied to a crop belongs to a different cross-resistance group than the last at-risk fungicide applied to the previous crop. Some at-risk fungicides are formulated as premix products with other fungicides to manage resistance. At-risk fungicides should be used only when needed most. The most critical time to use them for resistance management is early in an epidemic when the pathogen population is small. Multi-site contact fungicides should be used alone late in the growing season, where they have been shown to provide sufficient disease control to protect yield. Another important component of resistance management is assessing disease control and reporting any loss of efficacy potentially due to resistance.

2

PLANT PATHOLOGY AND PLANT PATHOGENS

A plant disease is defined as "anything that prevents a plant from performing to its maximum potential." This definition is broad and includes abiotic and biotic plant diseases.

ABIOTIC OR NON-INFECTIOUS DISEASES

These diseases are caused by conditions external to the plant, not living agents. They cannot spread from plant to plant, but are very common and should be considered when assessing the health of any plant. Examples of abiotic diseases include nutritional deficiencies, soil compaction, salt injury, ice, and sun scorch (Fig.1).

Fig. 1. Frost injury on soybean seedlings.

Fig. 2. Soybean plants dying from Sclerotinia infection.

BIOTIC OR INFECTIOUS DISEASES

These diseases are caused by living organisms. They are called plant pathogens when they infect plants. For the purposes of discussing plant pathology, only plant disease pathogens will be discussed. Pathogens can spread from plant to plant and may infect all types of plant tissue including leaves, shoots, stems, crowns, roots, tubers, fruit, seeds and vascular tissues (Fig. 2).

TYPES OF PLANT PATHOGENS

Plant pathogens are very similar to those that cause disease in humans and animals. Fungi, fungal-like organisms, bacteria, phytoplasmas, viruses, viroids, nematodes and parasitic higher plants are all plant pathogens.

Fungi and Fungal-like Organisms (FLOs)

Collectively, fungi and FLOs cause the most plant disease than any other group of plant pathogens. These organisms cannot make their own food, lack chlorophyll, have filamentous growth, and may or may not reproduce by spores. Fungi and FLOs are able to overwinter in soil or on plant debris. However, some fungi and FLOs cannot overwinter in northern climates because of low winter temperatures. These pathogens overwinter in southern climates and then are transported by air currents back to northern climates. Disease movement from southern to northern climates can be monitored during the growing season (Fig. 3).

Bacteria

Bacteria are single-celled microscopic organisms with cell walls that reproduce by binary fission (one cell splits into two). Introduction to the plant must occur through natural openings or wounds in the plant. Bacteria overwinter primarily in soil and in or on plant material that does not decompose, but some survive inside insect vectors (Fig. 4).

Fig. 3. Soybean infected with Sclerotinia.	**Fig. 4. Soybean infected by bacterial blight.**

Phytoplasmas

Phytoplasmasare microscopic, bacteria-like organisms that lack cell walls and thus appear filamentous (Fig. 5).

Fig. 5. Aster yellow phytoplasm infecting aster

VIRUSES AND VIROIDS

Viruses are intracellular (live inside the cell) nucleic acid particles with a protein coat that infect other living organisms and replicate in the hosts they infect. Viroids are virus-like particles but lack a protein coat. Viruses and viroids are primarily transmitted by vectors including insects, nematodes, and fungi, which introduce the virus or viroid during feeding. Viruses and viroids can also be transmitted through seed, vegetative propagation and pruning (Fig. 6).

NEMATODES

Nematodes are microscopic worm-like animals. The majority of nematodes are soil dwelling animals and move with soil. However, there are some nematodes that are transmitted through insects and infect above ground plant parts (Fig. 7).

Fig. 6.

Fig. 7. Adult soybean cyst nematode females emerging from soybean roots.

DISEASE TRIANGLE

Three components are absolutely necessary in order for a disease to occur in any plant system. The three components are:

1. a susceptible host plant
2. a virulent pathogen
3. a favorable environment

When these three components are present at the same time, a disease (shaded region) will occur if a susceptible host plant is in intimate association with a virulent plant pathogen under favorable environmental conditions. This concept is represented by the shaded portion of the diagram above. When there is a high degree of overlap (as the shaded area becomes larger), there will be a moderate to high amount of disease. (Fig. 8).

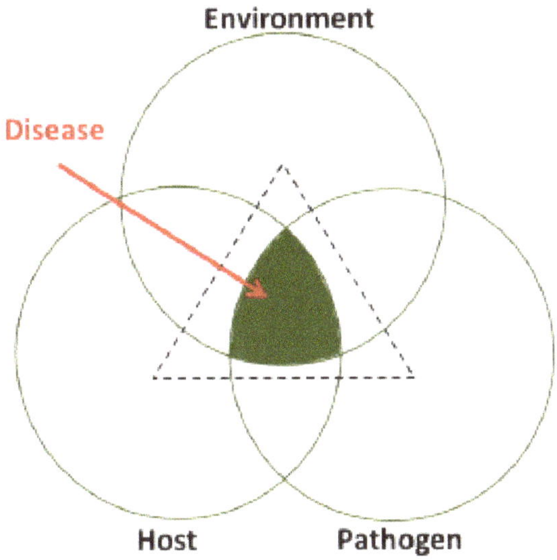

Fig. 8. Ven-diagram of disease triangle.

It is important to remember that within each of the three components -host, pathogen, and environment -there are numerous variables that may affect both the incidence and severity of the disease. These variables include genetic diversity, biology and lifecycle of the host plant and pathogen, and environmental conditions.

- Genetic diversity: Within one species of host plant there may be an incredible range of genetic diversity that greatly influences susceptibility to any particular species of pathogen. If the host is resistant to a pathogen, even when the pathogen is present under favorable environmental conditions, a disease will not occur (Fig. 9).Genetic diversity also plays a role in pathogen virulence or its ability to infect a host and cause disease, which may also influence the amount and severity of a disease.
- Biology and lifecycle of the host plant and pathogen: Host plants may be resistant to pathogens at one stage of development but not at another. In a similar manner, some pathogens must be at a critical life stage in order to cause infection.
- Environmental conditions: There are numerous variables in the environment that influence disease incidence and severity including temperature, sunlight, moisture, relative humidity, and time of year. Pathogens are typically restricted to an area based on the conditions of the macroclimate. A microclimate is the prevailing climatic conditions in a certain geographical area. Within a

macroclimate, small areas may exist in which the climate may be different than the surrounding areas. This is called a microclimate. Each landscape is filled with microclimates that exist because of differences in exposure to sun and wind, soil type and many other factors.

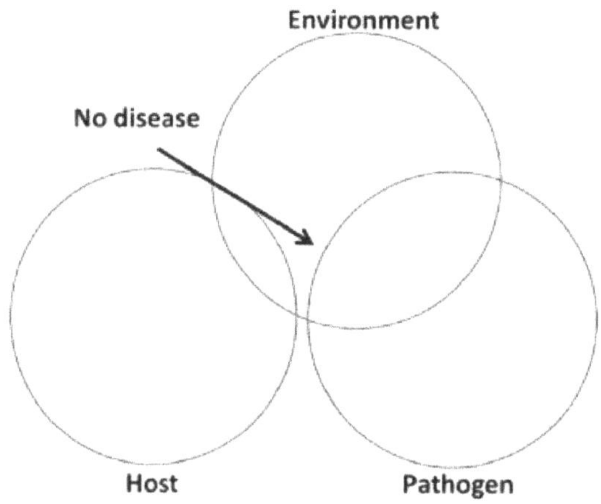

Fig. 9. Variables within each component of the disease triangle may affect the presence of disease.

BIOLOGICAL CONTROL OF PLANT PATHOGENS

Introduction

Plant diseases need to be controlled to maintain the quality and abundance of food, feed, and fiber produced by growers around the world. Different approaches may be used to prevent, mitigate or control plant diseases. Beyond good agronomic and horticultural practices, growers often rely heavily on chemical fertilizers and pesticides. Such inputs to agriculture have contributed significantly to the spectacular improvements in crop productivity and quality over the past 100 years. However, the environmental pollution caused by excessive use and misuse of agrochemicals, as well as fear-mongering by some opponents of pesticides, has led to considerable changes in people's attitudes towards the use of pesticides in agriculture.

Today, there are strict regulations on chemical pesticide use, and there is political pressure to remove the most hazardous chemicals from the market. Additionally, the spread of plant diseases in natural ecosystems may preclude successful application of chemicals, because of the scale to which such applications might have to be applied.

Consequently, some pest management researchers have focused their efforts on developing alternative inputs to synthetic chemicals for controlling pests and diseases. Among these alternatives are those referred to as biological controls.

A variety of biological controls are available for use, but further development and effective adoption will require a greater understanding of the complex interactions among plants, people, and the environment. To that end, this article is presented as an advanced survey of the nature and practice of biological control as it is applied to the suppression of plant diseases.

This survey will i) describe the various definitions and key mechanisms of biocontrol, ii) explore the relationships between microbial diversity and biological control, iii) describe the current status of research and application of biological controls, and iv) briefly outline future directions that might lead to the development of more diverse and effective biological controls for plant diseases.

DEFINITIONS

The terms "biological control" and its abbreviated synonym "biocontrol" have been used in different fields of biology, most notably entomology and plant pathology. In entomology, it has been used to describe the use of live predatory insects, entomopathogenic nematodes, or microbial pathogens to suppress populations of different pest insects. In plant pathology, the term applies to the use of microbial antagonists to suppress diseases as well as the use of host-specific pathogens to control weed populations. In both fields, the organism that suppresses the pest or pathogen is referred to as the biological control agent (BCA).

More broadly, the term biological control also has been applied to the use of the natural products extracted or fermented from various sources. These formulations may be very simple mixtures of natural ingredients with specific activities or complex mixtures with multiple effects on the host as well as the target pest or pathogen. And, while such inputs may mimic the activities of living organisms, non-living inputs should more properly be referred to as biopesticides or biofertilizers, depending on the primary benefit provided to the host plant. The various definitions offered in the scientific literature have sometimes caused confusion and controversy.

For example, members of the U.S. National Research Council took into account modern biotechnological developments and referred to biological control as "the use of natural or modified organisms, genes, or gene products, to reduce the effects of undesirable organisms and to favor desirable organisms such as crops, beneficial insects, and microorganisms", but this definition spurred much subsequent debate and it was frequently considered too broad by many scientists who worked in the field (US Congress, 1995). Because the term biological control can refer to a spectrum of ideas, it is important to stipulate the breadth of the term when it is applied to the review of any particular work.

Published definitions of biocontrol differ depending on the target of suppression; number, type and source of biological agents; and the degree and timing of human intervention. Most broadly, biological control is the suppression of damaging activities of one organism by one or more other organisms, often referred to as natural enemies.

With regards to plant diseases, suppression can be accomplished in many ways. If growers' activities are considered relevant, cultural practices such as the use of rotations and planting of disease resistant cultivars (whether naturally selected or genetically engineered) would be included in the definition. Because the plant host responds to numerous biological factors, both pathogenic and non-pathogenic, induced host resistance might be considered a form of biological control. More narrowly, biological control refers to the purposeful utilization of introduced or resident living organisms, other than disease resistant host plants, to suppress the activities and populations of one or more plant pathogens. This may involve the use of microbial inoculants to suppress a single type or class of plant diseases.

Or, this may involve managing soils to promote the combined activities of native soil- and plant-associated organisms that contribute to general suppression. Most narrowly, biological control refers to the suppression of a single pathogen (or pest), by a single antagonist, in a single cropping system. Most specialists in the field would concur with one of the narrower definitions presented above. In this review, biological control will be narrowly defined as highlighted above in bold.

Types of interactions contributing to biological control

Throughout their lifecycle, plants and pathogens interact with a wide variety of organisms. These interactions can significantly affect plant health in various ways. In order to understand the mechanisms of biological control, it is helpful to appreciate the different ways that organisms interact. Note, too, that in order to interact, organisms must have some form of direct or indirect contact. Odum (1953) proposed that the interactions of two populations be defined by the outcomes for each. The types of interactions were referred to as mutualism, protocooperation, commensalism, neutralism, competition, amensalism, parasitism, and predation. While the terminology was developed for macroecology, examples of all of these types of interactions can be found in the natural world at both the macroscopic and microscopic level. And, because the development of plant diseases involves both plants and microbes, the interactions that lead to biological control take place at multiple levels of scale.

From the plant's perspective, biological control can be considered a net positive result arising from a variety of specific and non-specific interactions. Using the spectrum of Odum's concepts, we can begin to classify and functionally delineate the diverse components of ecosystems that contribute to biocontrol. Mutualism is an association between two or more species where both species derive benefit. Sometimes, it is an obligatory lifelong interaction involving close physical and biochemical contact, such as those between plants and mycorrhizal fungi.

However, they are generally facultative and opportunistic. For example, bacteria in the genus Rhizobium can reproduce either in the soil or, to a much greater degree, through their mutualistic association with legume plants. These types of mutualism can contribute to biological control, by fortifying the plant with improved nutrition and/or by stimulating host defenses. Protocooperation is a form of mutualism, but the organisms involved do not depend exclusively on each other for survival. Many of the microbes isolated and classified as BCAs can be considered facultative mutualists involved in protocooperation, because survival rarely depends on any specific host and disease suppression will vary depending on the prevailing environmental conditions.

Further down the spectrum, commensalism is a symbiotic interaction between two living organisms, where one organism benefits and the other is neither harmed nor benefited. Most plant-associated microbes are assumed to be commensals with regards to the host plant, because their presence, individually or in total, rarely results in overtly positive or negative consequences to the plant. And, while their presence may present a variety of challenges to an infecting pathogen, an absence of measurable decrease in pathogen infection or disease severity is indicative of commensal interactions. Neutralism describes the biological interactions when the population density of one species has absolutely no effect whatsoever on the other.

Related to biological control, an inability to associate the population dynamics of pathogen with that of another organism would indicate neutralism. In contrast, antagonism between organisms results in a negative outcome for one or both. Competition within and between species results in decreased growth, activity and/or fecundity of the interacting organisms. Biocontrol can occur when non-pathogens compete with pathogens for nutrients in and around the host plant. Direct interactions that benefit one population at the expense of another also affect our understanding of biological control. Parasitism is a symbiosis in which two phylogenetically unrelated organisms coexist over a prolonged period of time.

In this type of association, one organism, usually the physically smaller of the two (called the parasite) benefits and the other (called the host) is harmed to some measurable extent. The activities of various hyperparasites, i.e., those agents that parasitize plant pathogens, can result in biocontrol. And, interestingly, host infection and parasitism by relatively avirulent pathogens may lead to biocontrol of more virulent pathogens through the stimulation of host defense systems. Lastly, predation refers to the hunting and killing of one organism by another for consumption and sustenance.

While the term predator typically refer to animals that feed at higher trophic levels in the macroscopic world, it has also been applied to the actions of microbes, e.g. protists, and mesofauna, e.g. fungal feeding nematodes and microarthropods, that consume pathogen biomass for sustenance. Biological control can result in varying degrees from all of these types of interactions, depending on the environmental context within which they occur. Significant biological control, as defined above, most generally arises from manipulating mutualisms between microbes and their plant hosts or from manipulating antagonisms between microbes and pathogens.

Mechanisms of biological control

Because biological control can result from many different types of interactions between organisms, researchers have focused on characterizing the mechanisms operating in different experimental situations. In all cases, pathogens are antagonized by the presence and activities of other organisms that they encounter. Here, we assert that the different mechanisms of antagonism occur across a spectrum of directionality related to the amount of interspecies contact and specificity of the interactions.

Direct antagonism results from physical contact and/or a high-degree of selectivity for the pathogen by the mechanism(s) expressed by the BCA(s). In such a scheme, hyperparasitism by obligate parasites of a plant pathogen would be considered the most direct type of antagonism because the activities of no other organism would be required to exert a suppressive effect. In contrast, indirect antagonisms result from activities that do not involve sensing or targeting a pathogen by the BCA(s).

Stimulation of plant host defense pathways by non-pathogenic BCAs is the most indirect form of antagonism. However, in the context of the natural environment, most described mechanisms of pathogen suppression will be modulated by the relative occurrence of other organisms in addition to the pathogen. While many investigations have attempted to establish the importance of specific mechanisms of biocontrol to particular pathosystems, all of the mechanisms described below are likely to be operating to some extent in all natural and managed ecosystems. And, the most effective BCAs studied to date appear to antagonize pathogens using multiple mechanisms.

For instance, pseudomonads known to produce the antibiotic 2,4-diacetylphloroglucinol (DAPG) may also induce host defenses. Additionally, DAPG-producers can aggressively colonize roots, a trait that might further contribute to their ability to suppress pathogen activity in the rhizosphere of wheat through competition for organic nutrients.

Table 1. Types of interspecies antagonisms leading to biological control of plant pathogens.

Type	Mechanism	Examples
Direct antagonism	Hyperparasitism/predation	Lytic/some nonlytic mycoviruses
		Ampelomyces quisqualis
		Lysobacter enzymogenes
		Pasteuria penetrans
		Trichoderma virens
Mixed-path antagonism	Antibiotics	2,4-diacetylphloroglucinol
		Phenazines
		Cyclic lipopeptides
	Lytic enzymes	Chitinases

Type	Mechanism	Examples
		Glucanases
		Proteases
	Unregulated waste products	Ammonia
		Carbon dioxide
		Hydrogen cyanide
	Physical/chemical interference	Blockage of soil pores
		Germination signals consumption
		Molecular cross-talk confused
Indirect antagonism	Competition	Exudates/leachates consumption
		Siderophore scavenging
		Physical niche occupation
	Induction of host resistance	Contact with fungal cell walls
		Detection of pathogen-associated, molecular patterns
		Phytohormone-mediated induction

Hyperparasites and predation

In hyperparasitism, the pathogen is directly attacked by a specific BCA that kills it or its propagules. In general, there are four major classes of hyperparasites: obligate bacterial pathogens, hypoviruses, facultative parasites, and predators. *Pasteuria penetrans* is an obligate bacterial pathogen of root-knot nematodes that has been used as a BCA.

Hypoviruses are hyperparasites. A classical example is the virus that infects *Cryphonectria parasitica*, a fungus causing chestnut blight, which causes hypovirulence, a reduction in disease-producing capacity of the pathogen. The phenomenon has controlled the chestnut blight in many places. However, the interaction of virus, fungus, tree, and environment determines the success or failure of hypovirulence.

There are several fungal parasites of plant pathogens, including those that attack sclerotia (e.g. *Coniothyrium minitans*) while others attack living hyphae (e.g. *Pythium oligandrum*). And, a single fungal pathogen can be attacked by multiple hyperparasites. For example, *Acremonium alternatum, Acrodontium crateriforme, Ampelomyces quisqualis, Cladosporium oxysporum,* and *Gliocladium virens* are just a few of the fungi that have the capacity to parasitize powdery mildew pathogens. Other hyperparasites attack plant-pathogenic nematodes during different stages of their life cycles (e.g. *Paecilomyces lilacinus* and *Dactylella oviparasitica*). In contrast to hyperparasitism, microbial predation is more general and pathogen non-specific and generally provides less predictable levels of disease control. Some BCAs exhibit predatory behavior under nutrient-limited conditions.

However, *Trichoderma* produce a range of enzymes that are directed against cell walls of fungi. However, when fresh bark is used in composts, *Trichoderma* spp. do not directly attack the plant pathogen, *Rhizoctonia solani*. But in decomposing bark, the concentration of readily available cellulose decreases and this activates the chitinase genes of *Trichoderma* spp., which in turn produce chitinase to parasitize *R. solani*.

Antibiotic-mediated suppression

Antibiotics are microbial toxins that can, at low concentrations, poison or kill other microorganisms. Most microbes produce and secrete one or more compounds with antibiotic activity. In some instances, antibiotics produced by microorganisms have been shown to be particularly effective at suppressing plant pathogens and the diseases they cause. Some examples of antibiotics reported to be involved in plant pathogen suppression are listed in Table 2.

In all cases, the antibiotics have been shown to be particularly effective at suppressing growth of the target pathogen *in vitro* and/or *in situ*. To be effective, antibiotics must be produced in sufficient quantities near the pathogen to result in a biocontrol effect. *In situ* production of antibiotics by several different biocontrol agents has been measured; however, the effective quantities are difficult to estimate because of the small quantities produced relative to the other, less toxic, organic compounds present in the phytosphere.

And while methods have been developed to ascertain when and where biocontrol agents may produce antibiotics, detecting expression in the infection court is difficult because of the heterogenous distribution of plant-associated microbes and the potential sites of infection. In a few cases, the relative importance of antibiotic production by biocontrol bacteria has been demonstrated, where one or more genes responsible for biosynthesis of the antibiotics have been manipulated.

For example, mutant strains incapable of producing phenazines or phloroglucinols have been shown to be equally capable of colonizing the rhizosphere but much less capable of suppressing soilborne root diseases than the corresponding wild-type and complemented mutant strains. Several biocontrol strains are known to produce multiple antibiotics which can suppress one or more pathogens. For example, *Bacillus cereus* strain UW85 is known to produce both zwittermycin and kanosamine.

The ability to produce multiple antibiotics probably helps to suppress diverse microbial competitors, some of which are likely to be plant pathogens. The ability to produce multiple classes of antibiotics, that differentially inhibit different pathogens, is likely to enhance biological control. More recently, *Pseudomonas putida* WCS358r strains genetically engineered to produce phenazine and DAPG displayed improved capacities to suppress plant diseases in field-grown wheat.

Antibiotic	Source	Target pathogen	Disease
2, 4-diacetyl-phloroglucinol	Pseudomonasfluorescens F113	Pythium spp.	Damping off
Agrocin 84	Agrobacterium radiobacter	Agrobacterium tumefaciens	Crown gall
Bacillomycin D	Bacillus subtilis AU195	Aspergillus flavus	Aflatoxin contamination
Bacillomycin, fengycin	Bacillus amyloliquefaciens FZB42	Fusarium oxysporum	Wilt
Xanthobaccin A	Lysobacter sp. strain SB-K88	Aphanomyces cochlioides	Damping off
Gliotoxin	Trichoderma virens	Rhizoctonia solani	Root rots
Herbicolin	Pantoea agglomerans C9-1	Erwinia amylovora	Fire blight
Iturin A	B. subtilis QST713	Botrytis cinerea and R. solani	Damping off
Mycosubtilin	B. subtilis BBG100	Pythium aphanidermatum	Damping off
Phenazines	P. fluorescens 2-79 and 30-84	Gaeumannomyces graminis var. tritici	Take-all
Pyoluteorin, pyrrolnitrin	P. fluorescens Pf-5	Pythium ultimum and R. solani	Damping off
Pyrrolnitrin, pseudane	Burkholderia cepacia	R. solani and Pyricularia oryzae	Damping off and rice blast
Zwittermicin A	Bacillus cereus UW85	Phytophthora medicaginis and	Damping off
		P. aphanidermatum	

Lytic enzymes and other byproducts of microbial life

Diverse microorganisms secrete and excrete other metabolites that can interfere with pathogen growth and/or activities. Many microorganisms produce and release lytic enzymes that can hydrolyze a wide variety of polymeric compounds, including chitin, proteins, cellulose, hemicellulose, and DNA. Expression and secretion of these enzymes by different microbes can sometimes result in the suppression of plant pathogen activities directly. For example, control of *Sclerotium rolfsii* by *Serratia marcescens* appeared to be mediated by chitinase expression.

And, a b-1,3-glucanase contributes significantly to biocontrol activities of *Lysobacter enzymogenes* strain C3. While they may stress and/or lyse cell walls of living organisms, these enzymes generally act to decompose plant residues and nonliving organic matter. Currently, it is unclear how much of the lytic enzyme activity that can be detected in the natural environment represents specific responses to microbe-microbe interactions.

It seems more likely that such activities are largely indicative of the need to degrade complex polymers in order to obtain carbon nutrition. Nonetheless, microbes that show a preference for colonizing and lysing plant pathogens might be classified as biocontrol agents. *Lysobacter* and *Myxobacteria* are known to produce copious amounts of lytic enzymes, and some isolates have been shown to be effective at suppressing fungal plant pathogens.

So, the lines between competition, hyperparasitism, and antibiosis are generally blurred. Furthermore, some products of lytic enzyme activity may contribute to indirect disease suppression. For example, oligosaccharides derived from fungal cell walls are known to be potent inducers of plant host defenses. Interestingly, *Lysobacter enzymogenes* strain C3 has been shown to induce plant host resistance to disease, though the precise activities leading to this induction are not entirely clear. The quantitative contribution of any and all of the above compounds to disease suppression is likely to be dependent on the composition and carbon to nitrogen ratio of the soil organic matter that serves as a food source for microbial populations in the soil and rhizosphere. However, such activities can be manipulated so as to result in greater disease suppression.

For example, in post-harvest disease control, addition of chitosan can stimulate microbial degradation of pathogens similar to that of an applied hyperparasite. Chitosan is a non-toxic and biodegradable polymer of beta-1,4-glucosamine produced from chitin by alkaline deacylation. Amendment of plant growth substratum with chitosan suppressed the root rot caused by *Fusarium oxysporum* f. sp. *radicis-lycopersici* in tomato. Although the exact mechanism of action of chitosan is not fully understood, it has been observed that treatment with chitosan increased resistance to pathogens.

Other microbial byproducts also may contribute to pathogen suppression. Hydrogen cyanide (HCN) effectively blocks the cytochrome oxidase pathway and is highly toxic to all aerobic microorganisms at picomolar concentrations. The production of HCN by certain fluorescent pseudomonads is believed to be involved in the suppression of root pathogens. *P. fluorescens* CHA0 produces antibiotics, siderophores and HCN, but suppression of black rot of tobacco caused by *Thielaviopsis basicola* appeared to be due primarily to HCN production. Researchers reported that volatile compounds such as ammonia produced by *Enterobacter cloacae* were involved in the suppression of *Pythium ultimum*-induced damping-off of cotton. While it is clear that biocontrol microbes can release many different compounds into their surrounding environment, the types and amounts produced in natural systems in the presence and absence of plant disease have not been well documented and this remains a frontier for discovery.

Competition

From a microbial perspective, soils and living plant surfaces are frequently nutrient limited environments. To successfully colonize the phytosphere, a microbe must effectively compete for the available nutrients. On plant surfaces, host-supplied nutrients include exudates, leachates, or senesced tissue. Additionally, nutrients can be obtained from waste products of other organisms such as insects (e.g. aphid honeydew on leaf surface) and the soil. While difficult to prove directly, much indirect evidence suggests that competition between pathogens and non-pathogens for nutrient resources is important for limiting disease incidence and severity.

In general, soilborne pathogens, such as species of Fusarium and Pythium, that infect through mycelial contact are more susceptible to competition from other soil- and plant-associated microbes than those pathogens that germinate directly on plant surfaces and infect through appressoria and infection pegs. Genetic work revealed that production of a particular plant glycoprotein called agglutinin was correlated with potential of *P. putida* to colonize the root system. *P. putida* mutants deficient in this ability exhibited reduced capacity to colonize the rhizosphere and a corresponding reduction in Fusarium wilt suppression in cucumber. The most abundant nonpathogenic plant-associated microbes are generally thought to protect the plant by rapid colonization and thereby exhausting the limited available substrates so that none are available for pathogens to grow.

For example, effective catabolism of nutrients in the spermosphere has been identified as a mechanism contributing to the suppression of *Pythium ultimum* by *Enterobacter cloacae*. At the same time, these microbes produce metabolites that suppress pathogens. These microbes colonize the sites where water and carbon-containing nutrients are most readily available, such as exit points of secondary roots, damaged epidermal cells, and nectaries and utilize the root mucilage.

Biocontrol based on competition for rare but essential micronutrients, such as iron, has also been examined. Iron is extremely limited in the rhizosphere, depending on soil pH. In highly oxidized and aerated soil, iron is present in ferric form, which is insoluble in water (pH 7.4) and the concentration may be as low as 10^{-18} M. This concentration is too low to support the growth of microorganisms, which generally need concentrations approaching 10^{-6} M. To survive in such an environment, organisms were found to secrete iron-binding ligands called siderophores having high affinity to sequester iron from the micro-environment.

Almost all microorganisms produce siderophores, of either the catechol type or hydroxamate type. Researchers were the first to demonstrate the importance of siderophore production as a mechanism of biological control of *Erwinia carotovora* by several plant-growth-promoting *Pseudomonas fluorescens* strains A1, BK1, TL3B1 and B10. And, a direct correlation was established in vitro between siderophore synthesis in fluorescent pseudomonads and their capacity to inhibit germination of chlamydospores of *F. oxysporum* (Elad and Baker 1985, Sneh et al. 1984).

As with the antibiotics, mutants incapable of producing some siderophores, such as pyoverdine, were reduced in their capacity to suppress different plant pathogens. The increased efficiency in iron uptake of the commensal microorganisms is thought to be a contributing factor to their ability to aggressively colonize plant roots and an aid to the displacement of the deleterious organisms from potential sites of infection.

Induction of host resistance

Plants actively respond to a variety of environmental stimuli, including gravity, light, temperature, physical stress, water and nutrient availability. Plants also respond to a variety of chemical stimuli produced by soil- and plant-associated microbes. Such stimuli can either induce or condition plant host defenses through biochemical changes that enhance resistance against subsequent infection by a variety of pathogens. Induction of host defenses can be local and/or systemic in nature, depending on the type, source, and amount of stimuli. Recently, phytopathologists have begun to characterize the determinants and pathways of induced resistance stimulated by biological control agents and other non-pathogenic microbes (Table 3).

Table 3. Bacterial determinants and types of host resistance induced by biocontrol agents

Bacterial strain	Plant species	Bacterial determinant	Type
Bacillus mycoides strain Bac J	Sugar beet	Peroxidase, chitinase and b-1,3-glucanase	ISR
Bacillus subtilis GB03 and IN937a	Arabidopsis	2,3-butanediol	ISR
Pseudomonas fluorescens strains			
CHA0	Tobacco	Siderophore	SAR
	Arabidopsis	Antibiotics (DAPG)	ISR
WCS374	Radish	Lipopolysaccharide	ISR
		Siderophore	
		Iron regulated factor	
WCS417	Carnation	Lipopolysaccharide	ISR
	Radish	Lipopolysaccharide	ISR
		Iron regulated factor	
	Arabidopsis	Lipopolysaccharide	ISR
	Tomato	Lipopolysaccharide	ISR
Pseudomonas putida strains	Arabidopsis	Lipopolysaccharide	ISR
WCS 358	Arabidopsis	Lipopolysaccharide	ISR
		Siderophore	ISR
BTP1	Bean	Z,3-hexenal	ISR
Serratia marcescens 90-166	Cucumber	Siderophore	ISR

The first of these pathways, termed systemic acquired resistance (SAR), is mediated by salicylic acid (SA), a compound which is frequently produced following pathogen infection and typically leads to the expression of pathogenesis-related (PR) proteins. These PR proteins include a variety of enzymes some of which may act directly to lyse invading cells, reinforce cell wall boundaries to resist infections, or induce localized cell death. A second phenotype, first referred to as induced systemic resistance (ISR), is mediated by jasmonic acid (JA) and/or ethylene, which are produced following applications of some nonpathogenic rhizobacteria. Interestingly, the SA- and JA- dependent defense pathways can be mutually antagonistic, and some bacterial pathogens take advantage of this to overcome the SAR.

For example, pathogenic strains of *Pseudomonassyringae* produce coronatine, which is similar to JA, to overcome the SA-mediated pathway. Because the various host-resistance pathways can be activated to varying degrees by different microbes and insect feeding, it is plausible that multiple stimuli are constantly being received and processed by the plant. Thus, the magnitude and duration of host defense induction will likely vary over time. Only if induction can be controlled, i.e. by overwhelming or synergistically interacting with endogenous signals, will host resistance be increased.

A number of strains of root-colonizing microbes have been identified as potential elicitors of plant host defenses. Some biocontrol strains of *Pseudomonas* sp. and *Trichoderma* sp. are known to strongly induce plant host defenses. In several instances, inoculations with plant-growth-promoting rhizobacteria (PGPR) were effective in controlling multiple diseases caused by different pathogens, including anthracnose (*Colletotrichum lagenarium*), angular leaf spot (*Pseudomonas syringae* pv. *lachrymans* and bacterial wilt (*Erwinia tracheiphila*). A number of chemical elicitors of SAR and ISR may be produced by the PGPR strains upon inoculation, including salicylic acid, siderophore, lipopolysaccharides, and 2,3-butanediol, and other volatile substances.

Again, there may be multiple functions to such molecules blurring the lines between direct and indirect antagonisms. More generally, a substantial number of microbial products have been identified as elicitors of host defenses, indicating that host defenses are likely stimulated continually over the course of a plant's lifecycle. Excluding the components directly related to pathogenesis, these inducers include lipopolysaccharides and flagellin from Gram-negative bacteria; cold shock proteins of diverse bacteria; transglutaminase, elicitins, and β-glucans in Oomycetes; invertase in yeast; chitin and ergosterol in all fungi; and xylanase in *Trichoderma*.

These data suggest that plants would detect the composition of their plant-associated microbial communities and respond to changes in the abundance, types, and localization of many different signals. The importance of such interactions is indicated by the fact that further induction of host resistance pathways, by chemical and microbiological inducers, is not always effective at improving plant health or productivity in the field.

Microbial diversity and disease suppression

Plants are surrounded by diverse types of mesofauna and microbial organisms, some of which can contribute to biological control of plant diseases. Microbes that contribute most to disease control are most likely those that could be classified competitive saprophytes, facultative plant symbionts and facultative hyperparasites. These can generally survive on dead plant material, but they are able to colonize and express biocontrol activities while growing on plant tissues. A few, like avirulent *Fusarium oxysporum* and binucleate

Rhizoctonia-like fungi, are phylogenetically very similar to plant pathogens but lack active virulence determinants for many of the plant hosts from which they can be recovered.

Others, like *Pythium oligandrum* are currently classified as distinct species. However, most are phylogenetically distinct from pathogens and, most often, they are subspecies variants of the same microbial groups. Due to the ease with which they can be cultured, most biocontrol research has focused on a limited number of bacterial (*Bacillus, Burkholderia, Lysobacter, Pantoea, Pseudomonas*, and *Streptomyces*) and fungal (*Ampelomyces, Coniothyrium, Dactylella, Gliocladium, Paecilomyces*, and *Trichoderma*) genera. Still, other microbes that are more recalcitrant to in vitro culturing have been intensively studied. These include mycorrhizal fungi, e.g. *Pisolithus* and *Glomus* spp. that can limit subsequent infections, and some hyperparasites of plant pathogens, e.g. *Pasteuria penetrans* which attack root-knot nematodes.

Because multiple infections can and do take place in field-grown plants, weakly virulent pathogens can contribute to the suppression of more virulent pathogens, via the induction of host defenses. Lastly, there are the many general micro- and meso-fauna predators, such as protists, collembola, mites, nematodes, annelids, and insect larvae whose activities can reduce pathogen biomass, but may also facilitate infection and/or stimulate plant host defenses by virtue of their own herbivorous activities.

While various epiphytes and endophytes may contribute to biological control, the ubiquity of mycorrhizae deserves special consideration. Mycorrhizae are formed as the result of mutualist symbioses between fungi and plants and occur on most plant species.

Because they are formed early in the development of the plants, they represent nearly ubiquitous root colonists that assist plants with the uptake of nutrients (especially phosphorus and micronutrients). The vesicular arbuscular mycorrhizal fungi (VAM, also known as arbuscular mycorrhizal or endomycorrhizal fungi) are all members of the zygomycota and the current classification contains one order, the Glomales, encompassing six genera into which 149 species have been classified.

Arbuscular mycorrhizae involve aseptate fungi and are named for characteristic structures like arbuscles and vesicles found in the root cortex. Arbuscules start to form by repeated dichotomous branching of fungal hyphae approximately two days after root penetration inside the root cortical cell. Arbuscules are believed to be the site of communication between the host and the fungus. Vesicles are basically hyphal swellings in the root cortex that contain lipids and cytoplasm and act as storage organ of VAM. These structures may present intra- and inter- cellular and can often develop thick walls in older roots. These thick walled structures may function as propagules.

During colonization, VAM fungi can prevent root infections by reducing the access sites and stimulating host defense. VAM fungi have been found to reduce the incidence of root-knot nematode. Various mechanisms also allow VAM fungi to increase a plant's stress tolerance. This includes the intricate network of fungal hyphae around the roots which block pathogen infections. Inoculation of apple-tree seedlings with the VAM fungi *Glomus fasciculatum* and *G. macrocarpum* suppressed apple replant disease caused by phytotoxic myxomycetes.

VAM fungi protect the host plant against root-infecting pathogenic bacteria. The damage due to *Pseudomonas syringae* on tomato may be significantly reduced when the plants are well colonized by mycorrhizae. The mechanisms involved in these interactions include physical protection, chemical interactions and indirect effects. The other mechanisms employed by VAM fungi to indirectly suppress plant pathogens include enhanced nutrition to plants; morphological changes in the root by increased lignification; changes in the chemical composition of the plant tissues like antifungal chitinase, isoflavonoids, etc.; alleviation of abiotic stress and changes in the microbial composition in the mycorrhizosphere.

In contrast to VAM fungi, ectomycorrhizae proliferate outside the root surface and form a sheath around the root by the combination of mass of root and hyphae called a mantle. Disease protection by ectomycorrhizal fungi may involve multiple mechanisms including antibiosis, synthesis of fungistatic compounds by plant roots in response to mycorrhizal infection and a physical barrier of the fungal mantle around the plant root. Ectomycorrhizal fungi like Paxillus involutus effectively controlled root rot caused by *Fusarium oxysporum* and *Fusarium moniliforme* in red pine. Inoculation of sand pine with *Pisolithus tinctorius*, another ectomycorrhizal fungus, controlled disease caused by *Phytophthora cinnamomi*.

Because plant diseases may be suppressed by the activities of one or more plant-associated microbes, researchers have attempted to characterize the organisms involved in biological control. Historically, this has been done primarily through isolation, characterization, and application of individual organisms. By design, this approach focuses on specific forms of disease suppression. Specific suppression results from the activities of one or just a few microbial antagonists. This type of suppression is thought to be occurring when inoculation of a biocontrol agent results in substantial levels of disease suppressiveness. Its occurrence in natural systems may also occur from time to time.

For example, the introduction of *Pseudomonas fluorescens* that produce the antibiotic 2,4-diacetylphloroglucinol can result in the suppression of various soilborne pathogens. However, specific agents must compete with other soil- and root-associated microbes to survive, propagate, and express their antagonistic potential during those times when the targeted pathogens pose an active threat to plant health. In contrast, general suppression is more frequently invoked to explain the reduced incidence or severity of plant diseases because the activities of multiple organisms can contribute to a reduction in disease pressure.

High soil organic matter supports a large and diverse mass of microbes resulting in the availability of fewer ecological niches for which a pathogen competes. The extent of general suppression will vary substantially depending on the quantity and quality of organic matter present in a soil. Functional redundancy within different microbial communities allows for rapid depletion of the available soil nutrient pool under a large variety of conditions, before the pathogens can utilize them to proliferate and cause disease.

For example, diverse seed-colonizing bacteria can consume nutrients that are released into the soil during germination thereby suppressing pathogen germination and growth. Manipulation of agricultural systems, through additions of composts, green manures and cover crops is aimed at improving endogenous levels of general suppression.

Biocontrol research, development, and adoption

Biological control really developed as an academic discipline during the 1970s and is now a mature science supported in both the public and private sector (Baker 1987). Research related to biological control is published in many different scientific journals, particularly those related to plant pathology and entomology. Additionally, three academic journals are specifically devoted to the discipline (i.e. Biological Control, Biocontrol Research and Technology, and BioControl). In the United States, research funds for the discipline are provided primarily by several USDA programs. These include the Section 406 programs, regional IPM grants, Integrated Organic Program, IR-4, and several programs funded as part of the National Research Initiative. Monies also exist to stimulate the development of commercial ventures through the small business innovation research (SBIR) programs. Such ventures are intended to be conduits for academic research that can be used to develop new companies.

Much has been learned from the biological control research conducted over the past forty years. But, in addition to learning the lessons of the past, biocontrol researchers need to look forward to define new and different questions, the answers to which will help facilitate new biocontrol technologies and applications. Currently, fundamental advances in computing, molecular biology, analytical chemistry, and statistics have led to new research aimed at characterizing the structure and functions of biocontrol agents, pathogens, and host plants at the molecular, cellular, organismal, and ecological levels..

Some of the research questions that will advance our understanding of biological controls and the conditions under which it can be most fruitfully applied are listed in Table 4.

Over the past fifty years, academic research has led to the development of a small but vital commercial sector that produces a number of biocontrol products. The current status of commercialization of biological control products has been reviewed recently.

As in most industries, funding in the private sector research and development goes through cycles, but seems likely to increase in the years ahead as regulatory and price pressures for agrochemical inputs increase. Most of the commercial production of biological control agents is handled by relatively small companies, such as Agraquest, BioWorks, Novozymes, Prophyta, Kemira Agro. Occasionally, such companies are absorbed by or act as subsidiaries of multi-billion dollar agrochemical companies, such as Bayer, Monsanto, Syngenta, and Sumitomo.

Total revenues of products used for biocontrol of plant diseases represented just a small fraction of the total pesticide market during the first few years of the 21st century with total sales on the order of $10 to 20 million dollars annually.

Table 4: Some current topics of biocontrol research and development and associated questions

1 The ecology of plant-associated microbes
How are pathogens and their antagonists distributed in the environment?
Under what conditions do biocontrol agents exert their suppressive capacities?
How do native and introduced populations respond to different management practices?
What determines successful colonization and expression of biocontrol traits?
What are the components and dynamics of plant host defense induction?
2 Application of current strains/inoculant strategies
Can more effective strains or strain variants be found for current applications?
Will genetic engineering of microbes and plants be useful for enhancing biocontrol?
How can formulations be used to enhance activities of known biocontrol agents?
3 Discovering novel strains and mechanisms of action
Can previously uncharacterized microbes act as biological control agents?
What other genes and gene products are involved in pathogen suppression?
Which novel strain combinations work more effectively than individual agents?
Which signal molecules of plant and microbial origin regulate the expression of biocontrol traits by different agents?
4 Practical integration into agricultural systems
Which production systems can most benefit from biocontrol for disease management?
Which biocontrol strategies best fit with other IPM system components?
Can effective biocontrol-cultivar combinations be developed by plant breeders?

However, significant expansion is expected over the next 10 years due to increasing petroleum prices, the expanded demand for organic food, and increased demand for "safer" pesticides in agriculture, forestry, and urban landscapes.

Growers are interested in reducing dependence on chemical inputs, so biological controls (defined in the narrow sense) can be expected to play an important role in Integrated Pest Management (IPM) systems.

A model describing the several steps required for a successful IPM has been developed. In this model, good cultural practices, including appropriate site selection, crop rotations, tillage, fertility and water management, provide the foundation for successful pest management by providing a fertile growing environment for the crop. The use of pest- and disease-resistant cultivars, developed through conventional breeding or genetic engineering, provides the next line of defense. However, such measures are not always sufficient to be productive or economically sustainable. In such cases, the next step would be to deploy biorational controls of insect pests and diseases

These include BCAs, introduced as inoculants or amendments, as well as active ingredients directly derived from natural origins and having a low impact on the environment and non-target organisms. If these foundational options are not sufficient to ensure plant health and/or economically sustainable production, then less specific and more harmful synthetic chemical toxins can be used to ensure productivity and profitability. With the growing interest in reducing chemical inputs, companies involved in the manufacturing and marketing of BCAs should experience continued growth.

However, stringent quality control measures must be adopted so that farmers get quality products. New, more effective and stable formulations also will need to be developed.

Most pathogens will be susceptible to one or more biocontrol strategies, but practical implementation on a commercial scale has been constrained by a number of factors. Cost, convenience, efficacy, and reliability of biological controls are important considerations, but only in relation to the alternative disease control strategies. Cultural practices (e.g. good sanitation, soil preparation, and water management) and host resistance can go a long way towards controlling many diseases, so biocontrol should be applied only when such agronomic practices are insufficient for effective disease control.

As long as petroleum is cheap and abundant, the cost and convenience of chemical pesticides will be difficult to surpass. However, if the infection court or target pathogen can be effectively colonized using inoculation, the ability of the living organism to reproduce could greatly reduce application costs. In general, though, regulatory and cultural concerns about the health and safety of specific classes of pesticides are the primary economic drivers promoting the adoption of biological control strategies in urban and rural landscapes.

Self-perpetuating biological controls (e.g. hypovirulence of the chestnut blight pathogen) are also needed for control of diseases in forested and rangeland ecosystems where high application rates over larger land areas are not economically-feasible. In terms of efficacy and reliability, the greatest successes in biological control have been achieved in situations where environmental conditions are most controlled or predictable and where biocontrol agents can preemptively colonize the infection court. Monocyclic, soilborne and post-harvest diseases have been controlled effectively by biological control agents that act as bioprotectants (i.e. preventing infections).

3

BOTANICAL FUNGICIDES IN AGRICULTURE

For many years, plant pathogenic fungi have caused devastating losses worth millions of dollars in crops worldwide. Synthetic fungicides have effectively controlled plant pathogenic fungi. Their repeated use over decades has disrupted natural biological systems, and sometimes resulted in development of fungal resistance. They had undesirable effects on non-target organisms, and fostered environmental and human health concerns. Therefore, alternative measures have been developed for crop protection. In an attempt to reduce the use of synthetic fungicides, intensive investigations into the possible exploitation of biopesticides such as natural commercial products that are safe for humans and the environment have been undertaken.

The global pesticide market was valued in 2009 at about \$42.8 billion. Of that total, nearly 3.74% is attributable to biopesticides. The global pesticide market is expected to expand further through 2014 for both synthetic pesticides and biopesticides. However, it is estimated that the use of synthetic insecticides will continuously decline, particularly in response to the momentum of organic agricultural practices, use of integrated pest management (IPM), and the development of genetically modified (GM) seeds with insecticidal qualities.

The synthetic pesticide market comprises herbicides, fungicides, insecticides, and others, accounting for 48%, 26%, 16%, and 10% of the total market, respectively. The pesticide market was declined over the period 1999 to 2004 because of the development of biopesticides and GM crops, especially herbicide-resistant seeds. However, in 2005, the use of herbicides began to increase, partly in response to the increased aggressiveness of herbicide-resistant weeds, the development of "super" weeds, and the expansion of sales to new markets.

Currently, the annual growth rate of synthetic pesticides is estimated to be 3% for the period 2009 to 2014. In contrast, the annual growth rate of bio-pesticides over the same period is expected to be 15.6%. The percentage of the total pesticide market for biopesticides is expected to increase to around 6% by 2014.

In USA, there are currently more than 245 registered biopesticide-active ingredients used in hundreds of products. Biopesticides are classified into microbial pesticides, biochemical biopesticides, plant incorporated protectants (PIPs; GM organisms or crops), predators, entomopathogenic nematodes, and parasitoids. Biopesticides account for about 20% of all pesticide-active ingredients registered in USA.

In Korea, a total of 33 biopesticides are registered, of which 19 are fungicides; 13, insecticides; 1, herbicide. Most of them are microbial pesticides; only 2 of 33 are botanical pesticides, which consists of 1 herbicide and 1 insecticide. No botanical fungicide has been registered in Korea.

In general, biopesticides are classified as natural products (microorganism-derived, higher plant-derived, and animal-derived products) and microorganisms (viruses, bacteria, fungi, nematodes, and protozoa). However, of the natural products, microorganism-derived compounds are not generally accepted as biochemical bio-pesticides in Korea, where they are recognized to be almost the same as synthetic pesticides. Therefore, many Korean scientists have been focused on the development of natural pesticides from plant-derived compounds. Research on botanical fungicides for agriculture has been intensified, since it has become evident that botanical fungicides have enormous potential to inspire and influence on modern agrochemical research.

The development of botanical fungicides would help decrease the negative impacts of synthetic agents, such as residues, resistance, and environmental pollution. In this respect, botanical fungicides may be effective, selective, biodegradable, and less toxic to the environment. In this paper, we reviewed commercially available plant-derived fungicides and various antifungal metabolites that show potent *in vivo* antifungal activities and have not yet been introduced into the market.

Cinnamaldehyde

Cinnamaldehyde (1, Fig. 1) was isolated from cinnamon essential oil in 1834 by Dumas and Peligot and synthesized in the laboratory by Chiozza in 1854. The chemical is usually synthesized for crop protection. It is effective against dry bubble caused by *Verticillium fungicola*, dollar spot caused by *Sclerotinia homeocarpa*, and pitch canker disease caused by *Fusarium moniliforme* var. *subglutinans*. Researchers reported antifungal activity of essential oils and their constituents from indigenous cinnamon (*Cinnamomum osmophloeum*) leaves against wood decay fungi.

Compared to the other components, cinnamaldehyde, the major compound in *C. osmophloeum* leaf essential oils, possessed the strongest antifungal activities and showed 100% efficacy against both *Coriolus versicolor* and *Laetiporus sulphureus*. Researchers reported the antifungal activity of cinnamaldehyde and eugenol congeners against wood-rot fungi. Results from antifungal tests revealed that cinnamaldehyde, alpha-methyl cinnamaldehyde, (*E*)-2-methylcinnamic acid, eugenol, and isoeugenol exhibited strong antifungal activity against all fungi tested. Moreover, researchers evaluated the effects of cinnamaldehyde, eugenol, citral, and geraniol on growth, hyphal ultrastructure, and virulence factors of *Aspergillus fumigates* and *Trichophyton rubrum*. The antifungal activity of their major constituents was evaluated in order cinnamaldehyde > eugenol > geraniol > citral.

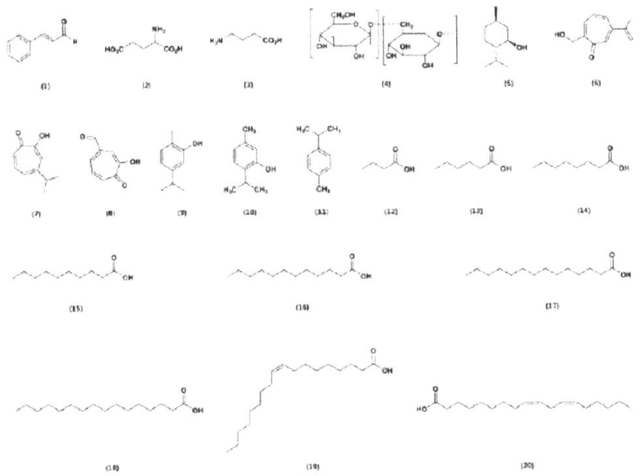

Fig. 1. Chemical structures of various plant-derived antifungal metabolites.

The mode of action of cinnamaldehyde is not understood. Its low toxicity and well-known properties make it ideal for agriculture. Cinnamaldehyde is sold as a 30% wettable powder (WP) formulation, which named Vertigo Wettable Powder Fungicide (Monterey), and cinnacure A3005 (Pro-guard). It is not soluble in water and is degraded rapidly in the soil and is not expected to pose any hazard to non-target organisms or the environment.

L-Glutamic acid + gamma-aminobutyric acid

Gamma-aminobutyric acid (GABA) (2) (29.2%) and L-glutamic acid (29.2%) are found in all living organisms. L-glutamic acid (3) is a major amino acid naturally found in plant and animal proteins, and GABA helps to maintain normal brain function. These two ingredients enhance the growth of specified plants such as almond, broccoli, and onions, prevent development of powdery mildew on grapes, and suppress certain other plant diseases such as brown rot (*Monilia* sp.) and shot hole (*Stigmina carpophila*) of stone fruit. The products are applied by spraying, drenching the soil, or via particular irrigation systems. Toxicity tests in animals and humans did not show any adverse effects from GABA or L-glutamic acid. Both substances occur naturally in plants and animals. Products containing the two substances are not approved for direct application to water or to areas where surface water is present.

Jojoba oil

Jojoba oil is a vegetable oil obtained from the jojoba bean. The physical properties of jojoba oil are similar to those of other vegetable oils. Jojoba has been found to be very effective and capable of controlling white flies and powdery mildew on ornamental plants and grapes. The ability of jojoba oil to remain stable even at high temperatures make it a widely usable fungicide in nearly all climatic conditions. One of its mechanisms of action is to form a physical barrier between an insect pest and the leaf surface. The fungicide is applied by ground application like a spray containing a final concentration of <1% jojoba oil.

Laminarin

Laminarin (also known as laminaran) (4) is a storage glucan (a polysaccharide of glucose) found in the blue green algae *Laminaria digitata*. Generally, □-glucans are major constituents in the bran of most cereal grains and intentionally added to many dietary supplements and texturing agents. It induces defense reactions in agricultural crops such as fruiting vegetables, including tomato, eggplant, pepper, zucchini, cucurbits, watermelon, melons, grape, apple, and pear. In particular, laminarin efficiently elicits defense responses in grapevine plants against gray mold and downy mildew caused by *Botrytis cinerea* and *Plasmopara viticola*, respectively. The development of these pathogens in infected grapevine plants was effectively suppressed by laminarin, and its application reduced infection by *B. cinerea* and *P. viticola* by approximately 55% and 75%, respectively. Researchers reported inhibitory effects of laminarin on the growth and toxin production of *A. flavus*. Significantly, inhibition of aflatoxin production in Sabouraud liquid medium was achieved using 150 μg/ml and 200 μg/ml of laminarin, without affecting mycelium growth. A natural fungicide containing laminarin is registered for use on wheat to control powdery mildew.

Milsana

Milsana is an ethanol extract from the plant *Reynoutria sachalinensis* for use on greenhouse-grown plants. Its application helps reduce certain plant diseases. Milsana is generally used as a preventative rather than a curative application, and the principal target disease is powdery mildew. Milsana controls powdery mildew on wheat by a combination of induced resistance and direct antifungal activity. A single spray to run-off, 48 h before inoculation, reduced mildew infection on young glasshouse seedlings by 97% (Reglinski, 2009). The active ingredient appears to be a natural elicitor of phytoalexins, which induce the plant's natural "immune system" and resistance in the host plant. Researchers reported the effects of *R. sachalinensis* extract and Milsana on conidial germination of *Leveillula taurica* and the efficacy of both Milsana formulations against the pathogen in the greenhouse. In addition, the research group reported field efficacy of 2 formulations of *R. sachalinensis* extract against powdery mildew (*Uncinula necator*) in grapes.

Pink plume poppy powder extract

Macleaya cordata is a bloodroot that has been used as an oriental medicine in China. Its extract contains sanguinaine and chelerythrine, which have fungicidal and insecticidal activities. Researchers reported that isoquinoline alkaloids from *M. cordata* were active against plant microbial pathogens. Among sanguinarine, chelerythrine, protopin, and alpha-allocryptopin isolated from *M. cordata*, sanguinarine displayed a significant antifungal activity against *Rhizoctonia solani* with a median inhibitory concentration (IC_{50}) of 0.45 μg/ml. Its crude extract and individual alkaloids are also active against several human pathogenic bacteria. The extract from *M. cordata* capsules had stronger antimicrobial effect than that from the aerial part of the plant or seeds and is sold as a 1.5% aqueous solution (Camas). Even though the mode of action has not been confirmed, the commercial product is known to induce systemic acquired resistance through accumulation of endogenous phenolic substances in the treated plants. This extract is a naturally occurring substance and is not considered to be toxic to non-target organisms, does not cause adverse environmental effects.

Essential oils

The essential oil from cymbopogan exhibited control values over all the plant and food rot mold tested. Researchers reported antifungal activities of essential oils of peppermint and sweet basil and their major aroma constituents in the vapor phase on some plant pathogenic fungi, such as *S. sclerotiorum*, *Rhizopus stolonifer*, and *Mucor* sp. Menthol (5) was found to be the individual aroma constituent responsible for the anti-fungal properties of peppermint essential oil, while menthone alone did not show any effect at the doses tested. In the case of basil oil, linalool alone showed a moderate antifungal activity while eugenol was completely inactive.

However, mixing the two components in a ratio similar to their concentrations in the original oil was found to enhance the antifungal properties of basil oil, indicating a synergistic effect. Researchers reported clear antifungal activity of β-dolabrin (6), β-thujaplicin (7), 4-acetyltropolone (8), and hinokitiol-related compounds against plant pathogenic fungi. In particular, β-dolabrin and 4-acetyltropolone showed strong antifungal activity against *Pythium aphanider-matum* IFO 32440, with minimum inhibitory concentration (MIC) at 6 μg/ml.

Researchers investigated inhibition of phytopathogenic fungi by essential oils, hydrosols, ground materials, and extract of summer savory (*Satureja hortensis*) growing wild in Turkey. The antifungal activities of all summer savory components on mycelial growth of *Alternaria mali* and *B. cinerea* were determined. All doses of extracts completely inhibited the mycelial growth of both fungi. Researchers reported that three essential oils extracted from *Cymbopogon citrates*, *Ocimum gratissimum*, and *Thymus vulgaris* are able to control seed-borne infection. The three essential oils applied as slurry controlled seed infection with a range of effectiveness of 48% to 100%. Researchers determined antifungal, phytotoxic, and insecticidal properties of the essential oil isolated from Turkish *Origanum acutidens* and its three components, carvacrol (9), thymol (10), and *p*-cymene (11).

The antifungal assays showed that *O. acutidens* oil, carvacrol, and thymol completely inhibited the mycelial growth of 17 phytopathogenic fungi, and their antifungal effects were greater than that of the commercial fungicide benomyl. However, *p*-cymene possessed lower antifungal activity. The oil, carvacrol, and thymol completely inhibited the seed germination and seedling growth of *Amaranthus retroflexus*, *Chenopodium album*, and *Rumex crispus* and also showed a potent phytotoxic effect against these plants. However, *p*-cymene did not show any phytotoxic effect. Furthermore, *O. acutidens* oil induced 68.3% and 36.7% mortality against *Sitophilus granarius* and *Trivolium confusum* adults, respectively.

Researchers reported antifungal activity against plant pathogenic fungi of essential oil and its constituents from *Calocedrus macrolepis* var. *formosana* Florin leaf. *T*-muurolol and alpha-cadinol strongly inhibited the growth of *R. solani* and *F. oxysporum*, with IC_{50} of less than 50 μg/ ml. These compounds also efficiently inhibited the mycelial growth of *Colletotrichum gloeosporioides*, *Pestalotiopsis funerea*, *Ganoderma australe*, and *F. solani*. Essential oils from 25 species of medicinal plants were tested as mycelial growth inhibitors against 6 important pathogenic and toxinogenic fungal species. All essential oils used in the agar dilution method evidently affected the growth of the 6 fungi.

Furthermore, researchers reported antifungal and herbicidal properties of essential oils and *n*-hexane extracts of *Achillea gypsicola* Hub-Mor. and *A. biebersteinii* (Asteraceae). The oils and hexane extracts were tested against 12 phytopathogenic fungi, and the oils were found to be more toxic than the hexane extracts. Of the fungi tested, mycelial growth of 11 fungi except *F. graminearum* was inhibited when treated with *A. gypsicola* oil. In addition, the extracts strongly increased the growth of *F. equiseti* and *F. graminearum*. Researchers screened antifungal activity of essential oils from Bergamot (*Citrus hystrix*) and tea tree (*Melaleuca alternifolia*) against economically important rice pathogenic fungi as *A. brassicicola*, *A. flavus*, *Bipolaris oryzae*, *F. moniliforme*, *F. proliferatum*, *M. oryzae* and *R. solani*. Tea tree oil at 2.0% v/v concentration showed the strongest inhibition of mycelial growth in *F. monoliforme*, *F. proliferratum*, *B. oryzae*, *A. brassicicola*, and *A. flavus*. Bergamot oil inhibited mycelial growth of all economically pathogenic fungi when applied at a 0.8% v/v concentration. However, it may not effectively control *Magnaporthe oryzae* and *R. solani*. Researchers reported that essential oils from *Asarum heterotropoides* var. *mandshuricum* inhibited the growth of the *A. humicola*, *C. gloeosporioides*, *R. solani*, and *Phytophthora cactorum*. These essential oils could be selected for developing lead compounds to control plant diseases.

Fatty acids

The potential of fatty acids in medical applications has been intensively investigated, but publications related to the biological activity of fatty acids against plant pathogens are relatively rare. Only few studies have been documented in this area: lauric acid and some unsaturated fatty acids have been shown to display inhibitory effects on several plant pathogens. Capric acid has been reported to possess the greatest cyto-toxicity against *Candida albicans*, and (*Z*)-9-heptadecenoic acid has an inhibitory effect against *Phytophthora infestans* and *Idriella bolleyi*.

Researchers reported antifungal activity of 10-hydroxystearic acid, 7*S*,10*S*-dihydroxy-8(*E*)-octadecenoic acid, and 12,13,17-trihydroxy-9(*Z*)-octadecenoic acid against *Blumeria graminis* f. sp. *tritici*, *Puccinia recondita*, *Pseudo-cercosporella herpotrichoides*, *Septoria nodorum*, *M. oryzae*, *R. solani*, *P. infestans*, and *B. cinerea*. Although 10-hydroxystearic acid and 7*S*,10*S*-dihydroxy-8(*E*)-octadecenoic acid showed no fungal disease control activity, 12,13,17-trihydroxy-9(*Z*)-octadecenoic acid did control plant pathogenic fungi, including *B. graminis*, *P. recondita*, *P. infestans*, and *B. cinerea*.

Researchers investigated antifungal activities of 9 fatty acids (butyric acid (12), caproic acid (13), caprylic acid (14), capric acid (15), lauric acid (16), myristic acid (17), palmitic acid (18), oleic acid (19), and linoleic acid (20) against 4 phytopathogenic fungi, *A. solani*, *C. lagenarium*, *F. oxysporum* f. sp. *cucumerinum*, and *F. oxysporum* f. sp. *lycopersici*. Pot experiments showed that a mixture of palmitic and oleic acids enhanced the growth of seedlings of continuous-tomato and continuous-cucumber. With the exception of oleic acid, in the petri dish assay, all fatty acids tested were observed to inhibit mycelial growth of one or more tested fungi. In particular, saturated fatty acids, e.g., palmitic acid, showed stronger antifungal activity than unsaturated fatty acids, e.g., oleic acid.

Two polyacetylenic acids, octadeca-9,11,13-triynoic acid (21) and *trans*-octadec-13-ene-9,11-diynoic acid (22), isolated from *Prunella vulgaris* inhibited the growth of various plant pathogens, including *M. oryzae*, *R. solani*, *P. infestans*, *S. sclerotiorum*, *F. oxysporum*, and *P. capsici*. In addition, these 2 compounds significantly suppressed the development of rice blast, tomato late blight, wheat leaf rust, and red pepper anthracnose, suggesting that fatty acids may be candidates for alternative approaches to the integrated control of phytopathogens.

Phenolic compounds

In recent years, a number of studies have been conducted on the antifungal activity of phenolic compounds found in medicinal plants. The compounds can be divided into simple phenolic compounds, flavones and related flavonoid glycosides, coumarins and derivatives, and anthraquinones.

Our research group reported *in vivo* antifungal activity against various plant pathogenic fungi of curcuminoids isolated from the rhizomes of *Curcuma longa*. Three curcuminoids, curcumin (23), demethoxycurcumin (24), and bisdemethoxy-curcumin (25), were purified from the methanol extract of *C. longa* rhizomes as antifungal agents. Among the three curcuminoids, demethoxycurcumin was the most active against both rice blast and tomato late blight, followed in order of curcumin and bisdemethoxycurcumin. In addition, the curcuminoids effectively inhibited the mycelial growth of three red pepper anthracnose pathogens, *C. coccodes*, *C. gloeosporioides*, and *C. acutatum* when applied in the range of 0.4–100 μg/ ml. The curcuminoids also highly inhibited spore germination in *C. coccodes*, with bisdemethoxycurcumin the most active in this respect. Among the three curcuminoids, only demethoxycurcumin was effective in suppressing red pepper anthracnose caused by *C. coccodes* in a greenhouse test. However, they all exhibited no or little *in vivo* antifungal activity against other fungal pathogens responsible for rice sheath blight, tomato gray mold, wheat leaf rust, and barley powdery mildew.

Our research group also reported antifungal activity of the malabaricones from *Myristica malabarica* fruit rinds against phytopathogenic fungi. Three antifungal resorcinols were identified: malabaricones A (MA; 26), B (MB; 27), and C (MC; 28). The three resorcinols inhibited mycelial growth in several plant pathogenic fungi, including *C. coccodes*, *C. gloeosporioides*, *M. oryzae*, and *R. solani*. The three compounds effectively reduced the development of rice blast, wheat leaf rust, and red pepper anthracnose. In addition, MC was highly active in reducing the development of tomato late blight.

Researchers investigated the antifungal activity of a kind of naphthoquinone, plumbagin (29), purified from leaves of *Nepenthes ventricosa* × *maxima* against phytopathogenic fungi, including *A. alternata*, *A. niger*, *B. oryzae*, *F. oxysporum*, *P. capsici*, *R. solani*, *R. stolonifer* var. *stolonifer*, and *S. sclerotiorum*. The compound exhibited antifungal activity against all the plant pathogenic fungi tested. The minimum inhibitory concentration values ranged from approximately 4.8 to 56.6 μg/ml against the 8 fungi listed above, with *R. solani* the most sensitive.

Chrysophanol (30), parietin (31), and nepodin (32) isolated from *R. crispus* rhizomes specifically reduced the development of barley and cucumber powdery mildews. The concentrations required for 50% disease control against barley powdery mildew caused by *B. graminis* f. sp. *hordei* were 4.7 μg/ml, 0.48 μg/ml, and 20 μg/ml for chrysophanol, parietin, and nepodin, respectively. Chrysophanol (100 μg/ml) and nepodin (400 μg/ ml) suppressed cucumber powdery mildew which is caused by *Podosphaera xanthii* more effectively than the fungicides fenarimol (30 μg/ml) and polyoxin B (100 μg/ml) under glasshouse conditions. Parietin (30 and 10 μg/ml) reduced the development of cucumber powdery mildew as efficiently as fenarimol (30 μg/ml) and more effectively than polyoxin B (100 μg/ml). Now several products containing chrysophanol and parietin are commercially available as ecofriendly organic agro-materials in Korea.

Dehydro–α–lapachone (33) isolated from *Catalpa ovata* stems showed a strong antifungal activity against plant pathogenic fungi. It completely inhibited the mycelial growth of *B. cinerea*, *C. acutatum*, *C. gloeosporioides*, *M. oryzae*, and *P. ultimum*. It also controlled the development of rice blast, tomato late blight, wheat leaf rust, barley powdery mildew, and red pepper anthracnose.

Decursin (34) and decursinol angelate (35) isolated from *Angelica gigas* effectively control rice blast via the inhibition of spore germination, rather than mycelial growth, in

M. oryzae (Yoon et al., 2011a). In addition, decursinol angelate reduced the development of wheat leaf rust and red pepper anthracnose. The three lignans erythro-austrobailignan-6 (EA6; 36), meso-dihydroguaiaretic acid (MDA; 37), and nectandrin-B (NB; 38) from *M. fragrans* displayed both antifungal and antibacterial activities. *In vitro* antimicrobial activity of the three lignans varied according to compound and target species. *A. alternata, C. coccodes, C. gloeosporioides, M. oryzae, Agrobacterium tumefaciens, Acidovorax konjaci,* and *Burk-holderia glumae* were relatively sensitive to the three lignans. *In vivo*, all three compounds effectively suppressed the development of rice blast and wheat leaf rust. In addition, EA6 and NB were highly active against the development of barley powdery mildew and tomato late blight, respectively. Both MDA and NB also moderately inhibited the development of rice sheath blight. Researchers reported effects of neolignans from the stem bark of *Magnolia obovata* on plant pathogenic fungi. Four neolignans from stem bark of *M. obovata* were identified as magnolol (39), honokiol (40), 4-methoxyhonokiol (41), and obovatol (42). Magnolol and honokiol showed similar *in vivo* antifungal spectra against 7 plant diseases tested; both compounds effectively suppressed the development of rice blast, tomato late blight, wheat leaf rust, and red pepper anthracnose. 4-Methoxyhonokiol and obovatol were highly active against only rice blast and wheat leaf rust, respectively.

Alkaloids

Pipernonaline (43), a piperidine alkaloid derived from long pepper (*Piper longum*), was reported to control the development of plant diseases caused by *M. oryzae, R. solani, B. cinerea, P. infestans, P. recondite,* and *B. graminis*. It showed a potent fungicidal activity against *P. recondite*, with 91% and 80% control values at concentrations of 0.5 μg/ml and 0.25 μg/ml, respectively.

Researchers reported that among fifiteen steroidal alkaloids from *Veratrum taliense* the verazine-type (veramitaline, stenophylline B, stenophyllin B-3-*O*–β–D-glucopyranoside, and veramiline-3-*O*–β–D-glucopyranoside) and jerveratrum-type (jervine, jervine-3-*O*–β–D-glucopyranoside) alkaloids exhibited strong antifungal activities against the phytopathogenic fungus *P. capsici*, with MICs of 160, 120, 160, 80, 80, and 120 μg/ml, respectively. Two alkaloids, D-calycanthin (44) and L-folicanthine (45), isolated from the seeds of *Chimonanthus praecox* showed significant inhibitory activity against five plant pathogenic fungi, *Ex-serohilum turcium, B. maydis, A. solani, S. sderotiorum,* and *F. oxysportium*. Singh et al. (2008) reported the antifungal activity of securinine (46) against some plant pathogenic fungi: *A. alternata, A. brassicae, A. brassicicola, Curvularia lunata, C. maculans, C. pallenscens, C. musae, Helminthosporium echinoclova,* and *H. spiciferum*.

Glycosides

Two sulfated triterpene glycosides, hemoiedemosides A (47) and B (48), from the Patagonian sea cucumber *Hemoiedema spectabilis* exhibited considerable antifungal activity against the phytopathogenic fungus *Cladosporium cucumerinum*. On the other hand, the semisynthetic desulfated hemoiedemoside A (47) was less active than two natural compounds hemoiedemosides A and B reported antifungal activity of triterpenic saponins from *Sapindus mukorossi* and *Diploknema butyracea* against phytopathogenic fungi. *D. butyracea* saponins exhibited significant antifungal activity against three fungi (ED_{50} 230–455 µg/ml), whereas *S. mukorossi* saponin was effective against two fungi (ED_{50} 181 – 407 µg/ml).

Pregnane glycosides from *Cynanchum wilfordii* are highly effective in reducing the development of powdery mildews (Yoon et al., 2011c). Among the six pregnane glycosides, three caudatin glycosides (49, 52, and 54) showed stronger antifungal activities than polyoxin B or the three kidjoranine glycosides (50, 51, and 53). A wettable powder-type formulation of the ethyl acetate extract from *C. wilfordii* roots prohibited the development of barley powdery mildew in growth chamber conditions and strawberry powdery mildew caused by *Sphaerotheca humuli* under greenhouse conditions. Our research group is currently developing a botanical fungicide for the environmentally benign control of powdery mildews.

4

DECISION-MAKING FACTORS FOR FUNGICIDE USE

Fungal disease management in U.S. corn production has undergone a major shift in the last 2 decades. Traditionally, corn producers managed fungal disease through biological and mechanical controls such as natural predators, crop rotations, hybrid seed selection, and tillage practices . In the mid-2000s, higher corn prices incentivized producers to shift to more continuous corn production and select hybrids based more on yield potential than disease resistance. These changes, coupled with less-intensive tillage practices and changes in weather patterns, have brought about more concerning fungal disease outbreaks.

Further, despite a lack of scientific consensus, some fungicide manufacturers started aggressive marketing campaigns suggesting fungicides provide yield benefits beyond controlling for fungal disease. The economics of fungal disease management changed, and many corn producers began to use fungicide (a chemical control) over traditional biological and mechanical controls.

Every year we hear about fungicides, and their utility in agronomic production systems. Indeed, these products can be an extremely useful tool to prevent yield-related losses resulting from fungal diseases.

However, reduced commodity prices are making it crucial for producers to minimize inputs in order to maximize profits. Fungicides can be near the top of the list when producers are deciding on what inputs to exclude during the season. Determining if an application is warranted can be a difficult decision to make.

Unfortunately this means I need to utilize the disease triangle or pyramid, which we all have seen innumerable times during discussions of plant diseases (Fig. 1). It does a great job of illustrating how diseases work, which is why it is utilized so often.

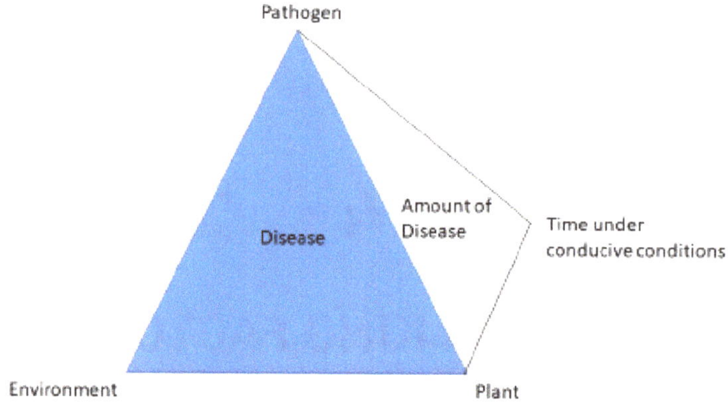

Fig. 1.The disease pyramid. Plant diseases only occur when the right pathogen is present with a susceptible plant and the environment is conducive for disease. The amount of time spent in disease favorable conditions results in the overall amount of disease.

THE DISEASE TRIANGLE (PYRAMID)

- Plant diseases are caused by pathogens. In order for a disease to occur, the correct pathogen (species, race, pathotype) must be present, and at the correct growth stage to infect the plant.
- A given pathogen can only infect a susceptible host plant. This means that the plant needs to be the correct species, lack resistance to the pathogen, and be at the correct growth stage required by the pathogen for infection. Different pathogens may infect different hosts and have different host ranges.
- Disease only occurs under the correct environment. This means that the environment is conducive for the particular disease (temperature, moisture, light). Different pathogens have different environmental requirements for growth and infection.
- The amount of disease is impacted by the amount of time all three points in the disease triangle are met. The longer all the disease requirements are met, the worse the disease, and the greater impacts it may have on crop productivity.

It is also important to note that different crops can handle different levels of disease before yield is impacted. The benefits of making a fungicide application are most pronounced, and you are most likely to see a return on your investment, when it prevents the disease from reaching the economic threshold. This is the level of disease where making a fungicide application will provide an economic return (Fig. 2).

Fungicides help to take part of the disease triangle out of the equation, assuming the other parts of the triangle (host, environment) are or will be met within 2-3 weeks of application. If a yield-limiting fungal disease epidemic occurs, the application of a fungicide would help reduce the epidemic and control disease-related losses.

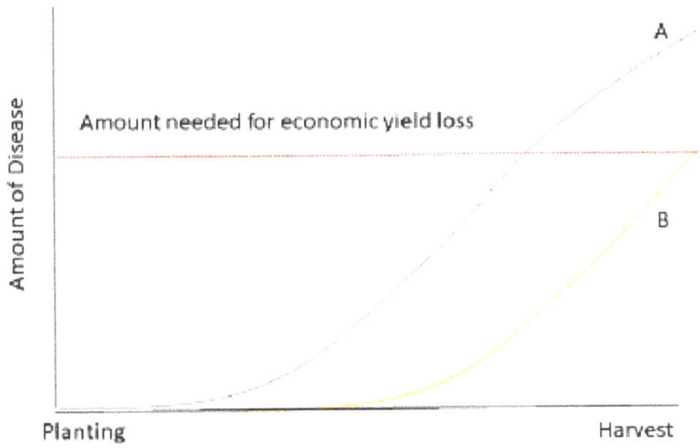

Fig. 2. Examples of how a fungicide can help reduce yield related losses. A) An untreated field with conditions favoring the development of a foliar fungal disease that causes economic yield loss. B) The application of a fungicide pushes back disease progress, preventing or minimizing economic yield loss.

Let's boil all that down to a couple brief statements:

- Fungal diseases only occur when the correct pathogen, plant host, and environment occur together
- The amount of disease is related to the duration of time conditions for disease are met
- Not all disease results in yield loss
- Fungicides minimize fungal disease
- Economic returns are greatest when fungicides are applied to susceptible plants and conditions favoring disease occur.

Thus, fungicides can minimize disease related yield loss. Any "yield increases or bumps" are related to the level of disease that would have occurred if the fungicide was not applied. Yes, there are some fungicides that can have subtle impacts on plant physiology. Sometimes these effects can have small impacts on plant yield. However, these physiological effects are highly unpredictable. When a fungicide is applied to control a disease, the result is highly predictable, given that a disease occurs. This should be the 1# reason for using a fungicide-reducing the impacts of fungal diseases to mimize yield losses.

REASONS FOR MAKING A FUNGICIDE APPLICATION DECISION:

Making a fungicide application decision needs to take in consideration the types of pathogens most often encountered in the field/location, the resistance rating of the hybrid or cultivar you are using to the pathogens most likely to be encountered, and the management practices you implement on that field to help reduce disease over time.

Remember that many diseases such as grey leaf spot on corn or frogeye leaf spot on soybean, are predominantly residue borne. Any practices that help reduce local inoculum buildup, for example, crop rotation and tillage, helps to degrade that residue and reduce pathogen populations, such that the amount of potential disease is reduced. Utilizing these practices helps to push the curve further to the right, in the case of Figure 2, resulting in a population that takes longer to reach a level where yield loss would occur. If you do not rotate and use no till or minimal tillage systems, that residue remains on the surface and these pathogens can build up.

During the season, other factors can increase risk. For example, heavy rains can promote disease. In the case of diseases that are not residue borne, but blow in from other regions, such as Southern rust on corn, the presence of the disease in nearby regions may increase risk. Perhaps the environment has been very conducive to disease, and disease has started early in the season? How high are you on the risk ladder?

This leads me to my last point regarding making fungicide decisions: conduct your own research. With all the money you are putting into your crop, it makes sense to invest a little extra time to determine if your decision was a good one. That knowledge can be used to help you make more profitable decisions in subsequent seasons.

There are many different ways you can do on farm trials, and they can range from extremely simple to complex. Even a simple test can provide useful information. Regardless of if you want to do replicated strips and calculate statistics, or simply want to split a field, it is important to always include an untreated strip. Without an untreated strip, *you cannot draw conclusions if you do not see a difference*. You also have no idea of the amount of yield that was protected by the application of the fungicide.

For example, if you want to look at two products and split the field in half, what happens if you see no yield difference between the field sections? Is it because the products are equal or because there wasn't enough disease to cause a yield reduction? A single untreated check can provide you a basis for making conclusions regarding the application of a fungicide to a particular field.

When you include a strip, make sure it runs through a representative part of the field. Do not locate it near field edges or unrepresentative parts of the field. Make sure the strip is wide enough so you can determine yield. The more tests you run with an untreated strip, the more confidant you will be in making fungicide application decisions in the future.

FOLIAR FUNGICIDE APPLICATION TIMING FOR MANAGING LEAF RUST

Flutriafol in-furrow (400mL/ha) applied across the site at seeding by the grower protected the crop from significant powdery mildew and rust up until crop heading (Z55). Initially in early August (at Z39) there was a trace of powdery mildew low in the canopy but it was gone in September. Leaf rust was first detected in crop on 16 August 2016 when the crop was at Z55, a few leaf rust pustules were found on the outer edges of the paddock that the trial was in but not easily found in the trial area itself. There were two leaf rust pathotypes detected (104-1,3,4,6,7,8,10,12+Lr37 and 76-1,3,5,7,9,10,12 +Lr37) both of which have increased virulence in Mace, as evidenced by the level of infection in untreated plots and subsequent yield impact.

Leaf spot diseases (yellow spot and septoria nodorum blotch) progress shows that the best fungicide application in this trial for yellow spot/septoria nodorum blotch diseases was the historically accepted optimal timing of Z39. Spraying at two (Z55) and four (Z65) weeks after flag leaf emergence gave less effective disease control, as infection had already occurred while leaves were unprotected.

For leaf rust on the other hand, this trial showed it was more important to apply fungicide when rust was first detected in the area rather than delay for two weeks or rely on earlier growth stage timings. The only significant yield response in the trial was from the foliar fungicide applied at time of rust detection Z55 (4.7t/ha compared to untreated which was 4.3t/ha). Fungicide applied at this time was able to protect leaves as disease was developing, while the protection from the earlier spray had diminished by this time and the later spray allowed too much infection to establish before spraying.

Factors to consider before applying foliar fungicide

Factors to consider before applying foliar fungicides include:

- The disease pressure – dependent on variety, weather events and inoculum source
- Whether there are visual symptoms present – monitor the crop for infection levels
- Whether you are applying the fungicide as a seed treatment, in-furrow or foliar
- The growth stage of your crop – apply fungicides to protect plant parts that contribute most to yield
- The crop yield potential – in low disease + low yielding crops, fewer sprays are needed. In moderate to heavy disease + higher yielding crops, multiple fungicide inputs are often required.

What is the objective of applying a foliar fungicide?

The main goal of a foliar fungicide application is to increase profitability.

Fungicides do not generally increase yield, but they protect the yield potential of a given crop by maintaining the green vegetative area (including leaves, stems and seed pods).

Foliar diseases reduce the green area of the plant and its ability to produce energy (via photosynthesis), particularly during key growth stages such as pod fill.

Getting the best result

There are a number of key considerations when applying fungicides to achieve the best result. These factors include:

- Timing of application
- Product selection
- Rate selection.

Timing of application

Timing of application of foliar fungicides is the most important consideration for effective disease control.

Key points to consider with fungicide timing are:

- Application at critical growth stages - protect plant tissues that contribute to yield
- Multiple applications may be required as fungicides cannot protect leaves that haven't emerged at the time of spraying
- High disease pressure/susceptible variety = earlier spraying + multiple sprays
- Applying fungicides to prevent the disease from becoming established. Do not rely on curative ability.

Product selection

The choice of fungicide to apply will depend on:

- The disease/s present
- Severity of disease pressure
- Crop growth stage
- Yield potential
- Resistance management
- Likelihood of livestock grazing.

Fungicides differ in their disease spectrum and efficacy. Mathcing the right product to the pathogen at the right timing provides a greater likelihood of getting a profitable return on investment.

Rate Selection

Changing the application rate of foliar fungicides will alter the efficacy, disease spectrum and length of residual control.

Growers may choose to apply lower rates under lower disease pressure and/or when multiple applications are planned in a spray program.

If higher rates are to be used, growers must ensure they do not exceed the maximum number of sprays permitted for the season or the maximum combined product rate allowable per hectare as directed by the label or permit. This is to ensure there are not excessive residues in the grain or hay, as well as ensuring fungicide resistance management is considered.

FACTORS IN THE SUCCESS OR FAILURE OF FOLIAR FUNGICIDE APPLICATION

Effects of foliar fungicide on yield, micronutrients

Wheat (*Triticum aestivum* L.) is currently grown on a greater land area than any other crop. In 2018 alone, over 214 million hectares were harvested worldwide, with the United States as the fourth-largest producer, only behind India, Russia, and China. In the United States, the Great Plains, which include the Dakotas, Nebraska, Kansas, Oklahoma, and parts of Montana, Wyoming, Colorado, New Mexico, and Texas, are a major wheat-producing region, accounting for the %30 and %50 of the annual production of all and winter wheat, respectively.

Plant breeding has offered a valuable tool to meet the demand for wheat from the constantly growing global population and face the challenges posed by climate change. Since the Green Revolution starting in the early 1960s, plant breeders have successfully improved the yield of wheat through enhanced agronomic practices (e.g., fungicide application) and by selecting for desirable traits such as short stature, adaptation to abiotic stress, and disease resistance.

Plant breeders estimate genetic yield gains over time by simultaneously evaluating modern and old cultivars in their adapted environments in replicated and randomized field trials. Upon harvest, linear regression analyses are generally performed to assess significant trends in yield and other variables over time. In this manner, the productivity of hard winter wheat cultivars adapted to the Great Plains of the United States has been estimated to improve at an annualized rate of 10 to 18 kg ha^{-1} yr^{-1} due to plant breeding efforts.

Nevertheless, these genetic gains of grain yield might negatively affect the nutritional quality of wheat. For instance, the grain protein concentration (GPC) of wheat in the Great Plains has been shown to decrease by 10 mg kg^{-1} for every kg ha^{-1} increase in grain yield. Therefore, high yielding modern cultivars generally contain less protein than historical varieties.

Besides serving as a primary source of calories, wheat contributes to the dietary intake of essential micronutrients for populations that rely on it as a staple crop. The inadequate intake of micronutrients such as iron (Fe) and zinc (Zn) leads to health complications such as anemia and stunting, respectively. The concentrations of Fe, Zn, copper (Cu), and magnesium (Mg) have decreased in modern British wheat cultivars when compared to older cultivars. In the Great Plains of the U.S., decreasing trends of Fe and Zn in grains of modern wheat varieties have also been reported. Reports on the decrease in micronutrient concentrations in modern compared with historical wheats have driven attention towards older varieties as these are recognized for their higher seed nutrition. Nevertheless, studies addressing the impact of the application of fungicide, as a standard agronomic practice, on micronutrients in historical and modern wheats are scarce.

Because of its higher concentration of mineral elements, dietary fiber, and potential health benefits, the consumption of whole wheat flour has gained popularity. Nevertheless, antinutritional compounds in the bran of wheat might attenuate some of these health benefits. Phytic acid, the main storage form of phosphorus (P) in grains, concentrates in the aleurone layer and forms insoluble complexes with mineral elements, collectively referred to as phytate, thus decreasing their bioavailability.

Indeed, the low phytic acid trait is associated with increased bioavailability of mineral elements in wheat flour. In one report, this antinutritional compound steadily increased during the last eighty years of crop development in Pakistan ; however, there are no studies on the temporal trends of phytate concentration in cultivars derived from the last century of plant breeding in the Great Plains.

Besides its potentially lower nutritional quality, modern wheat might serve as a vehicle of toxic heavy metals. Cadmium (Cd), a heavy metal recognized as a human carcinogen, is introduced into cropland as a result of industrial and agronomic activities. Since wheat plants inadvertently accumulate Cd in grains, wheat is recognized as a primary source of dietary Cd for the general population. Multiple governments have established maximum permissible levels of Cd in grains (0.1–0.2 mg kg^{-1}) to prevent detrimental health effects such as bone demineralization, tumors, and kidney dysfunction.

There is one study on the variability of Cd in a panel of historical and modern hard winter wheat from the Great Plains germplasm. Although high heritability was found for Cd, the relationships between registration year and the concentration of grain Cd was not evaluated.

Extensive research on the dynamics of Cd in durum wheat have led to the reduction of grain Cd concentrations in modern cultivars. The discovery of the gene *Cdu1*, controlling the root-to-shoot Cd translocation in durum wheat, has allowed breeders to introduce the low-Cd trait in modern varieties. Additionally, several studies have shown a negative correlation between plant height and grain Cd. As the leaves and stems compete as Cd sinks during grain filling, less Cd reaches the grains in taller wheat cultivars. Nevertheless, the molecular basis of this relationship is yet to be determined.

The understanding of the dynamics of Cd in bread wheat is in a more premature state than that in durum wheat. A homoeologous locus to *Cdu1* in bread wheat largely explaining the grain Cd variation across cultivars has not been identified. Although negative correlations between plant height and grain Cd have also been reported, only an infrequent reduced height gene (Rht8) has been associated to grain Cd. In a genetic mapping study, a wheat recombinant inbred line (RIL) population (n = 105) segregating for grain Cd concentration levels was used to identify alleles with increased Cd concentrations. Results revealed that RILs possessing the GWM261 allele on chromosome (Chr) 5A (conditioning shorter parenchyma cells in internode and peduncle) and the heavy metal accumulating *HMA3* allele (conditioning elevated Cd concentrations) also on Chr 5A had the highest Cd concentrations among all RILs evaluated in the study, with additive effects for elevated Cd levels. The GWM261 locus is tightly linked (0.6 cM) to the reduced height *Rht8* locus. Therefore, breeders using the *Rht8* locus to select for reduced height and resistance to lodging were advised to concurrently select for low-Cd alleles at the *HMA3*-linked locus on Chr 5A to develop new dwarfing varieties with lower seed Cd concentration.

In addition to plant breeding, agronomic practices, such as the application of fungicides, could also affect the concentration of micronutrients and Cd accumulation in wheat grains. On the one hand, fungicide could decrease mineral element concentrations because of increased yields. On the other hand, higher concentrations of elements could occur if the application of fungicide extends the grain filling stage, thereby enabling more mineral elements to be deposited into the grains.

The objectives of the present study were to estimate genetic gains and historical trends for important economic (yield and GPC) and nutritional (grain mineral elements, phytate, and Cd) traits of hard winter wheat by evaluating a historic set of eighteen wheat cultivars and two landraces adapted for production in the Great Plains in the presence and absence of fungicide treatment.

PLANT MATERIALS AND EXPERIMENTAL DESIGN

These cultivars, originating from the U.S. states of Kansas, Nebraska, Texas, and Oklahoma, were selected based on their historical relevance for grain production, known adaptation of climate conditions of the location of the study, and their contribution to the pedigrees of modern genotypes widely grown in the Great Plains. Nineteen out of the twenty genotypes evaluated herein are present in the hard winter association mapping panel, and their genotyping data is publicly available. The two landraces were included as references of the improvements of elite cultivars due to plant breeding. Because among the elite cultivars only 'Wesley' was found to carry the dwarfing *Rht8* allele related to grain Cd, comparisons among genotypes based on their plant height were not pursued.

For each of two growing seasons, 2017 (2016–2017) and 2018 (2017–2018), the experimental design consisted of a split-plot design with main plots in randomized complete blocks of three replications. The whole plot treatments were two levels of fungicide (treated and untreated). A single application of a broad-spectrum fungicide (Caramba, BASF, NJ, USA) was performed at the booting stage at the dose rate recommended by the manufacturer to simulate the common agronomic practice of the region. The twenty cultivars were planted in 1 m^2 plots within each block as the split-plot treatments.

In 2017, each plot was visually evaluated for foliar disease severity two weeks after fungicide intervention. To report plant health, the flag leaf was scored from 0 to 10, where

0 represented a highly damaged leaf and 10 a completely healthy green leaf. 2018 data were not recorded, thus are not reported in this manuscript. The grain yield (kg ha^{-1}) for each cultivar was determined by harvesting the full plot with a simple plot combine harvester. The diameter and thousand kernels dry weight (TKW) of the grains were determined using a single-kernel characterization system 4100 (Petern Instruments, Huddinge, Sweden).

This research was carried out at the University of Nebraska Eastern Nebraska Research and Extension Center (ENREC) near Mead, NE USA. In an effort to create measurable variability in the grain Cd concentration among the samples, this location was selected because of its known elevated concentrations of soil Cd. Depending on soil properties such as organic matter content and pH, the soil diethylene triamine pentaacetic acid (DTPA) extractable Cd at this location was reported to range between 0.16 to 0.26 mg kg^{-1}, whereas 0.04 mg kg^{-1} have been reported for surrounding locations. The high heavy metal concentration of this land is thought to be due to its use by the U.S. military for munitions manufacturing during the World War II era.

MINERAL ELEMENT ANALYSIS

Mineral elements were determined from whole wheat kernels using a wet ashing procedure followed by quantification through inductively coupled plasma mass spectrometry. In brief, 2 g of wheat kernels were washed with 0.3 N hydrochloric acid in a Buchner funnel and then rinsed with distilled water to remove debris and adhering elements from the surface of the kernels. The samples were then predigested overnight in 3.5 mL of concentrated nitric acid at room temperature and then digested by incubating for 1 h at 40°C followed by 1 h at 100°C in a BD50 digestion block (Seal-Analytical, WI, USA). Subsequently, 3.5 mL of nitric acid were added, and the tubes were incubated for another 1 h at 100°C. Next, 4 mL of 30% hydrogen peroxide were added and the samples allowed to incubate at 125°C for 1.5 h followed by another dose of 4 mL of hydrogen peroxide and an additional 1 h at 125°C. The samples were then evaporated to dryness by raising the temperature to 150°C for 2 h and then 165°C until the sample appeared dry.

The dried ash was resuspended in 12 mL of 1% nitric acid containing 50 ppb gallium as an internal standard and mineral elements were analyzed using a inductively coupled plasma mass spectrometer (7500cx ICP-MS Agilent Technologies Inc., CA, USA) operating in kinetic discrimination mode with helium gas at 5 mL min^{-1}. The injection volume was 40 ¼L; each sample replicate was measured in duplicate, and 2% nitric acid was used to rinse between samples. The results obtained for Cd, Fe, Mg, Mn, P, and Zn are presented in this research.

WHEAT MILLING, PROTEIN, AND PHYTATE CONCENTRATION OF WHOLE WHEAT FLOUR

Prior to milling, wheat kernels were dried to approximately 8% moisture content in a convection oven at 40°C overnight. The milling was performed using a laboratory roller mill (Quadrumat Jr, CW Brabender Instruments Inc., South Hackensack, NJ, USA) with the sifting screen removed to recover the resulting whole wheat flour. For the determination of protein and phytate concentrations, the flour was manually mixed before sampling to evenly distribute the bran particles.

The GPC of whole wheat flours was measured using a nitrogen analyzer (FP 528, LECO, St. Joseph, MI, USA) with a protein conversion factor of 5.7.

Phytate was measured as previously reported. In brief, phytic acid was extracted from 250 mg of whole wheat flour in 10 mL of 0.2 N hydrochloric acid by shaking overnight in a 4°C water bath. The contents were centrifuged (3000 g, 5 min), the supernatant diluted to a final volume of 25 mL with distilled water, and the pellet discarded. Next, 0.25 mL of the extract was mixed with 0.75 mL of 0.2 N hydrochloric acid, and 1.0 mL of 450 ¼M ammonium iron (III) sulfate and placed in a boiling water bath for 30 min. Subsequently, the tubes were cooled and centrifuged (3000 g, 5 min).

One mL of the supernatant was mixed with 1.5 mL of 2,22 -bipyridine, and the absorbance was measured at 530 nm. A standard curve of sodium phytate dodecahydrate was used to quantify the concentration of phytate in the samples.

STATISTICAL ANALYSES

The data from both growing years were analyzed using a three-factor factorial ANOVA. Registration year (as a continuous variable), fungicide, growing year, and their interactions were treated as fixed effects, whereas replications nested within growing year was treated as a random effect. Because of the significant effect of growing year on multiple response variables, data were analyzed within each individual growing year to evaluate the effects of registration year and fungicide in each growing year. A correlation analysis between the data from each growing year was performed to evaluate whether the independent variables followed similar temporal trends in both growing years despite differences in their magnitude.

To visualize the relationships among the independent variables and registration year, all variables were adjusted for growing year to account for its magnitude effect. In brief, each independent variable was regressed on growing year (as continuous variable) and the residuals saved. To obtain values in a practical range, the mean of the raw data was added to the residuals. Finally, these adjusted variables were regressed on registration year.

To evaluate the dilution effect of the increasing yield on the historical trends of the response variables, a partial correlation analysis, accounting for the effect of growing year and yield, was performed for variables that showed significant regressions with registration year. As mentioned above, in this work the landraces 'Turkey' (1874) and 'Kharkof' (1900) were included as references of the improvement of elite cultivars due to plant breeding. However, these landraces were not included in the regression analyses due to the outlying nature of their release years in the context of the period studied (1933–2013).

A principal component analysis was performed to determine the influence of yield on the effect of fungicide on grain micronutrient concentrations. This data analysis was performed using R-studio (version 1.2.5019–6).

Environmental conditions differed between the two growing years, but economic and nutritional traits of wheat presented similar temporal dynamics in both growing years

First, the weather conditions from the 2017 and 2018 growing seasons differed considerably. 2017 was considered a good year in terms of yield in the region. Dry conditions allowed for planting during early October. Due to the wet conditions in April and May, stripe rust (*Puccinia stritiformis* Westend) and powdery mildew (*Blumeria graminis* f. sp. *tritici*) were the major diseases observed in 2017, whereas the warm dry conditions that followed during flowering from late May to mid-June prevented fusarium head blight (*Fusarium graminearum*) during this growing season.

On the other hand, 2018 was an abnormal year. High precipitation in early October delayed planting until the last week of the month. The unusually dry and cold weather in November, combined with late planting, reduced tillering and negatively affected yield. The region also experienced the coldest April and warmest May on record. Due to low precipitation, there was little disease pressure during the spring.

However, a rainy June allowed for higher disease pressure after flowering with leaf rust (*Puccina triticina*), bacterial leaf streak (*Xanthomonas translucens*), and fusarium head blight (*Fusarium graminearum*) as the main diseases observed.

Growing year, as a main effect or interacting with fungicide or registration year, had a significant effect on all the economical and nutritional traits studied. This is likely due to the considerable differences in environmental conditions between the two growing years, as discussed. Nevertheless, the positive correlations between all the response variables across the two growing years, except for Mg in the presence of fungicide, indicated that despite the significant effect of growing year, similar trends were observed in both growing seasons under each fungicide treatment.

This suggests changes in magnitude of the effects due to environment rather than crossover effects. To account for this magnitude effect in the analysis of fungicide treatment, the response variables were partially regressed on registration years, while controlling for the effect of growing year under each fungicide treatment for further analysis.

DIFFERENTIAL TEMPORAL DYNAMICS AND BENEFITS OF FUNGICIDE ON THE ECONOMIC TRAITS OF WHEAT

The improvement in yield over time is known as the genetic gain obtained through plant breeding and may also be expressed as a percentage of the yield of a check cultivar or historic landrace grown in the region. In the case of the Great Plains, 'Kharkof' is commonly used as the check cultivar. Therefore, the rates obtained in this study are equivalent to 0.86% and 0.47% yr^{-1} over Kharkof in the presence and absence of fungicide since 1933, respectively.

The fungicide treatment presented an interaction with registration year, yet, in general, the application of fungicide increased the mean yield of all cultivars. The range of yield increased from 3101–4418 kg ha^{-1} to 5794–3685 kg ha^{-1} under the fungicide treatment since 1933. While the landraces Turkey and Kharkof increased from 3094 to 3564 kg ha^{-1} and from 2738 to 3089 kg ha^{-1}, respectively.

In 2017, which was characterized by a higher foliar disease pressure during spring, as discussed above, the disease severity was decreased by fungicide in all cultivars, as evidenced by increases in leaf health, except for Kharkof (1900) and 'Lancer' (1963). The amount of leaf health improvement differed for all cultivars. Those genotypes with high disease severity such as 'Turkey' (1874), 'Wichita' (1944), 'Warrior' (1960), 'Triumph 64' (1964), 'Centurk 78' (1978), 'Centura' (1983), 'TAM 107' (1984), and 'Settler CL' (2009) had greater benefits from the treatment than less susceptible cultivars.

All the cultivars presented similar grain sizes, including the landraces. Registration year presented non-significant regressions with grain diameter or TKW. Whereas, fungicide application led to significantly bigger and heavier kernels with means increasing from 2.6 to 2.7 mm of diameter and the TKW from 26 to 28 g.

A three-way interaction between registration year, fungicide, and growing year was significant for GPC. In general, GPC decreased over the period studied. The concentration of protein decreased at an annualized rate of 190 and 180 mg kg^{-1} yr^{-1} under the application and absence of fungicide, respectively. The partial correlations between GPC and registration year, when controlling for growing year and yield, remained significant in the presence (r = 032-., $P < 0.001$) and absence of fungicide (r = 0.30-, $P < 0.01$). The effect of fungicide on GPC was inconsistent, as it presented a significant interaction with growing year, which was only observed in 2018.

GRAIN CADMIUM INCREASED OVER TIME IN THE ABSENCE OF FUNGICIDE WHILE ITS APPLICATION ATTENUATED THIS TREND

Grain Cd increased with registration year in the absence of fungicide (Fig 3). Most samples, regardless of fungicide treatment, were within 0.1 and 0.2 mg kg^{-1}, the maximum permissible limits in China and Europe, respectively. The concentration of grain Cd significantly increased at an annualized rates of 0.4 ¼g kg^{-1} yr^{-1} in the absence of fungicide (P <0.01). The application of fungicide decreased the range of grain Cd concentrations from 0.1–0.19 mg kg^{-1} to 0.14–0.07 mg $kg^{-1,}$ while no significant temporal trend was observed under fungicide application (Fig 3).

PHYTATE AND TOTAL PHOSPHORUS REMAINED CONSTANT OVER TIME WHILE FUNGICIDE HAD CONTRASTING EFFECTS ON THESE COMPONENTS

When controlling for the effect of growing year, the grain phytate concentration remained constant across the registration years studied (Fig 4A).

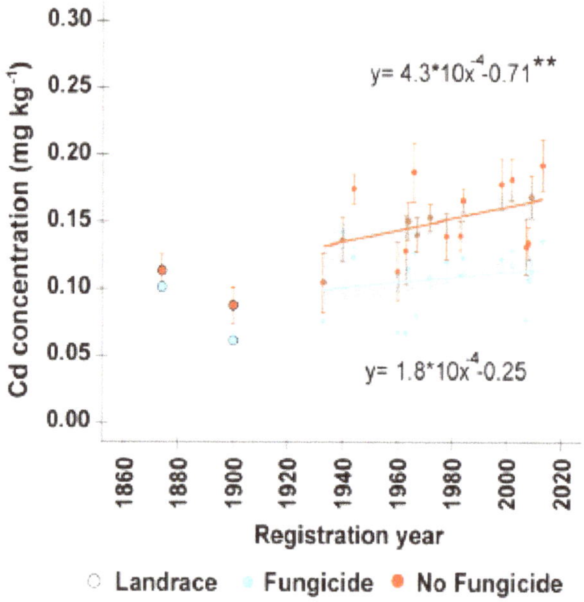

Fig 3. Linear regression of grain cadmium (Cd) concentration by registration year for a historic set of 18 wheat cultivars or landraces released between 1933 and 2013

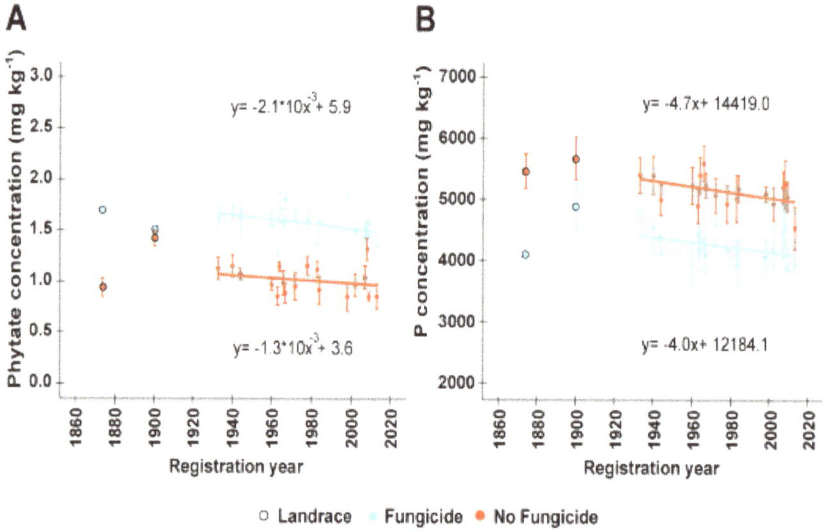

Fig 4. Linear regressions of (A) phytate and (B) phosphorus (P) concentrations by registration year for a historic set of 18 wheat cultivars or landraces released between 1933 and 2013

CONDUCTING ON-FARM COMPARISONS TO TEST FUNGICIDES

As farmers and farm advisers, you make many management choices during a growing season. For example, you may have to decide which cultivar to plant, which herbicide to use, how frequently to apply a fungicide, and what rate of nematicide to use. Often the information needed to make the best decision is available to you, but when it is not available you can frequently compare the options by conducting your own small experiments. Your experiments can be just as valid as any university study if you follow a few important principles of experimental design.

All experiments have certain things in common, so designing an experiment usually includes the following steps. You must decide what question you want to have answered. This is the goal, or objective, of the experiment. The goal of the experiment will dictate what to include in the experiment to help you answer your question. The individual things that you wish to test in your experiment are called "treatments" and the physical areas to which the treatments are applied are called "plots." Then you need to decide how the treatments should be physically arranged in the field. Technically, this is what is called the "experimental design."

Experiments answer your original question by allowing you to make unbiased comparisons among the treatments you selected. You will need some way to evaluate how well each treatment worked to make comparisons among treatments. The information you collect to help you make those comparisons (such as yield, insect counts, or disease severity) is called "data." Finally, you need an objective way to evaluate the data. This is usually done through statistical analysis.

SELECTING TREATMENTS

The objective, or purpose, of the study will determine the treatments included in an experiment. Writing down the test objectives is helpful because this forces you to define them precisely. A test may have more than one objective, although multiple objectives should be closely related.

The selection of treatments is usually logical if you can define the purpose of the study; all treatments necessary to address the test's objective should be included. For example, if the purpose were to determine which of five fungicides works the best, then the treatments would include all five of those fungicides. If the purpose were to determine if any of the five fungicides works better than your current choice, then the treatments would include the five fungicides plus the fungicide you currently use.

Accurately stating the purpose of the test before the treatments are applied in the field is critical. After the treatments have begun, it will be too late to add other treatments to answer the question you really wanted to address.

The selection of treatments and the experimental design get more complicated as the question you are trying to answer gets more complex. It is common to want to test in the same experiment two (or more) things that influence crop production. For example, you may want to test chicken litter as a fertilizer and test five corn hybrids to maximize yield. The specific questions addressed in this case are:

1. What effect does chicken litter have on corn production?
2. What effect do the hybrids have on corn production?
3. Does chicken litter have the same effect on each hybrid?

The third question may not be as obvious as the first two, but it will always be asked or implied if you are testing two or more factors in the same experiment. In this example, you have to determine what the effect of chicken litter is on each hybrid and then compare those effects to each other. To do that, the treatment list must include each hybrid without chicken litter and each hybrid with chicken litter (a total of 10 treatments). With this list of treatments, you can make the comparisons necessary to answer our three questions. This example employs a "factorial arrangement of treatments" that will be discussed in more detail in a later section.

One final note about treatment selection involves including additional treatments to provide a relative measure of effect. Comparing the yield of five new corn hybrids does little good if you cannot tell how those yields compare with the hybrids you already grow. You should include at least one hybrid with which you are already familiar (often called a "standard" treatment) to provide a relative measure of how well the new hybrids produce. If you wish to test a new nematicide, you should include a treatment with the currently used nematicide and a treatment with no nematicide as a basis for comparison. Such treatments are known as "checks" or "controls." Without the proper controls, you will not be able to say that the new nematicide worked better than the currently used nematicide or even that the new nematicide worked better than no nematicide! The questions you wish the experiment to answer should indicate what treatments should be included as controls.

It is frequently desirable to have both a positive and a negative control in an experiment. The negative control helps you determine if the treatments being tested work better than some minimal treatment (or nothing) and positive controls help you determine if the treatments being tested work better than the current standard practice. You may have several control treatments in an experiment if you currently have several viable options from which to choose. For example, if you currently can choose either of two fungicides to control a leafspot problem, you may wish to include them both as controls in your experiment when you test new products. You do not have to include all currently available options as controls for the experiment to be useful, but you can.

REPLICATION

In an experiment, replication means that individual treatments (such as each of the five pesticides being tested in an experiment) have been applied to more than one plot. Replication is necessary because all test plots are not identical, and that leads to variation in the data you collect; you will not get exactly the same results from two plots that received the same treatment. You can take steps to minimize the effect of variation if it has an identifiable cause, but there will always be some variation among plots that cannot be controlled. In statistical terms, uncontrolled variation is called experimental error. The purpose of replication is to allow you to make a more accurate estimate of how each treatment performed even though there is uncontrolled variation in the experiment. This can best be shown in an example.

Suppose you have 10 rose bushes and you want to test if a new fungicide will protect the bushes from black spot, a fungal leaf disease. You could pick five plants to leave untreated as a control and spray the other five with the fungicide. Later, when black spot is evident on the leaves, you count the number of diseased spots on each plant and compare the two treatments. The five untreated plants have 26, 21, 19, 25, and 23 infected spots (a treatment mean, or average, of 22.8 spots per plant), and the fungicide treated plants have 20, 15, 18, 21, and 20 spots (a mean of 18.8). Statistical analysis indicates that the fungicide did in fact reduce the number of infected sites, but you would not be able to determine that if you only had one treated and one untreated plot. Think about the following examples based on the data above. If you only had two plants and the untreated plant had 26 spots and the treated plant had 15 spots, it may seem easy to determine that the fungicide reduced disease. But what if the untreated plant had 19 spots and the treated plant had 21 spots? You might conclude that the fungicide did not work or even that it increased disease! Adequate replication can minimize this problem.

It is common when you have several replications of each treatment to have data like that in the rose example above: The treatment means are different but individual measurements may overlap. In this example, the lowest measurement from the untreated plants was 19, and the highest measurement from the fungicide treated plants was 21, but the treatment means were 22.8 for the untreated plants and 18.8 for the treated plants. Replication of treatments increases your ability to detect differences in treatment means. Having more replications allows you to identify (statistically) smaller differences in treatment means than you could identify with fewer replications.

The number of replications that you need is influenced by the biology of what you are testing, how close together the treatment means are, and how much variation exists within a treatment. For field tests in plant pathology, nematology, weed science, soil fertility studies, and entomology, a minimum of four replications is suggested, but five or six replications are much better. If treatment means are close together or variation is relatively large among the plots that received the same treatment, then you may need more replications to detect differences among treatments.

Just as the data may vary within a replicated treatment, the results may vary among experiments if the whole experiment is repeated. This can happen because of different weather conditions, different disease or insect pressure, or many other factors beyond your control. This does not mean that the results of a single experiment are not valid, but it does make it dangerous to draw conclusions from a single experiment. The one set of results you have may indicate treatment differences, but if you repeated the test several times you might not see those treatment differences again. If the test is repeated (and that means you cannot change any of the treatments) and you get similar results, then you can be much more confident that your conclusions are correct.

RANDOMIZATION

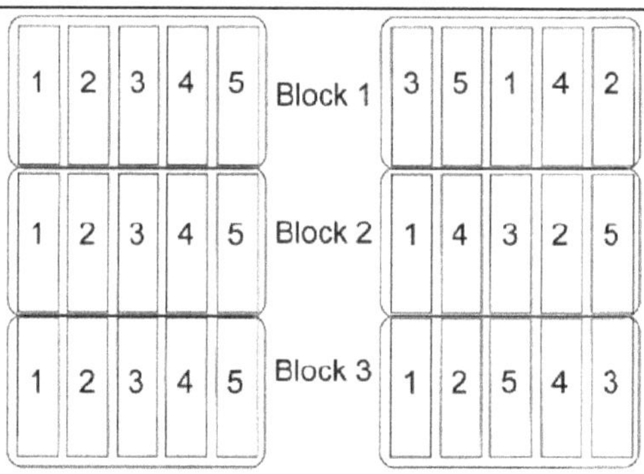

Fig. 5. Test plots on the left are not randomized. Plots on the right are randomized. The numbers (1-5) represent the five treatments in this test.

Randomization in an experiment means that the treatments are assigned to plots with no discernable pattern to the assignments. The reason randomization is important is that the positioning of treatments within the block may affect their performance. One example of this is an experiment testing five corn hybrids (labeled 1 through 5) in which you plant the hybrids in the same order in each block: 1, 2, 3, 4, then 5 (see Fig 5). If hybrid 2 is naturally much taller than the others, it can slightly shade the hybrids planted next to it (hybrids 1 and 3) and unfairly make them look a little bit worse than they would look if they were not planted next to hybrid 2. Another example is a field in which soil fertility gets progressively lower as you cross the field from east to west so that productivity is reduced as you go from one side of the field to the other. If two corn hybrids are planted side by side but within a block, hybrid 1 is always planted on the east side of hybrid 2, then hybrid 1 is always planted in slightly more fertile soil and therefore has an unfair advantage.

In both of the preceding examples, randomization could have prevented the unintentional bias because the arrangement of the treatments would have been different within each block. Because you cannot anticipate all the influences that may introduce bias into a test, ALL experiments should be randomized. There are many ways to randomize treatments within a block, but the simplest is literally to pull the numbers out of a hat.

Assign each treatment a number, write the numbers on individual pieces of paper, mix the slips of paper up, and then select the slips one at a time without looking at them first. The order in which the numbers are drawn is the order in which they will be arranged in a block. Repeat these steps for each block in the experiment.

If you wish to use the experiment as a demonstration plot (such as for a field day), it is common that one block not be truly randomized. This is done so that particular treatments can be seen side-by-side to facilitate comparisons and highlight differences for casual observers. Though it is better to randomize all blocks and not intentionally arrange treatments, arranging the treatments in one block is unlikely to affect the test's results as long as the other blocks are truly randomized.

PLOT SIZE

A plot, the area to which an individual treatment is applied, can be any size, including a single plant growing in a pot or 5 acres or more of a field. Before you can apply treatments to your test area, you must decide how large your plots should be. Although there is a lot of subjectivity in selecting plot size, there are some important considerations including the equipment to be used in planting, harvesting, and treatment application; how much space is available for use in the experiment and how many other treatments there are; and the biology of what you are studying. Accommodating equipment and space concerns makes it easier to conduct the test. Accommodating biological concerns reduces the chances of overlooking differences among treatments. Equipment and space considerations are usually easy to identify, but biological considerations are not always obvious.

If you have equipment to plant, harvest, and apply treatments to four rows at a time, then the logical plot width would be some multiple of four rows (four rows, eight rows, 12 rows, etc.). Any other width (such as six rows) would make it more difficult to conduct the experiment. The length that plots should be is more flexible than plot width. For example, if you plan to weigh the harvest from each plot, the scales you have may influence the length plots should be. If you have scales that are designed for weighing hundreds of pounds, your plots should be large enough to provide a harvest weight that can be accurately determined by your equipment, and increasing the length of plots is an easy way to do that. Also, the length of your plots may be adjusted so that all of your plots (all replications of all treatments) will fit into the area available for your test. If you have a large area for your test, space may not be an important consideration.

To accommodate biological considerations, you should answer two questions:

1. How large a plot is needed to observe the biological effect (disease severity, insect damage, weed frequency, nematode population levels, etc.) that you are studying?
2. How large a plot is needed to minimize the influence of a treatment (chemical application, etc.) on the plots next to it?

By answering these questions, you can determine the minimum plot size necessary to get useful data from the experiment.

To get an accurate measurement of the effect of pest management treatments, the plot must be large enough to account for uneven initial distribution of the pest (pathogen, insect, weed, etc.). Some areas may start with the pest present, but the pest may occur in other areas only after it has spread from its initial location. This is very important for pests that spread very slowly (such as most soilborne organisms).

Some diseases and pests are highly mobile and spread very rapidly (such as many insects). In an insect management trial, measuring the effect of a treatment can be very difficult if your plots are too small because the insects that you see in the plot may have simply spread from the plot next to it. To minimize this problem, you can increase your plot size and then collect data from the middle section of the plot. For example, you might have an eight-row plot but only collect data from the middle four rows. The rows from which you do not collect data are often referred to as "buffer rows" because they buffer the effect of the neighboring plots. If you do not use buffer rows when they are needed, you may fail to detect differences among treatments and incorrectly conclude that many treatments were ineffective. Buffer rows are frequently used when there is uncertainty whether treatments can influence nearby rows.

A similar concept involves the use of border rows along the edges of your test area. A significant "border effect" commonly exists at the edge of a field where the plants may grow differently than plants not at the edge. Although you may be able to minimize this problem with blocking, it is often better to eliminate the problem by not using the rows at the edge of a field in your experiment.

Once the plots are large enough to be representative of a much larger area, further increasing plot size will not significantly improve the accuracy of the results. For example, in an experiment testing fungicides for control of white mold, or stem rot, in peanut, a four-row-wide by 100-foot-long plot should be just as good as an eight-row-wide by 400-foot-long plot. Plots that are larger than necessary take more field space and may increase the amount of work required for an experiment, but they usually will not adversely affect the test results unless the plots get so large that the plots within a block are no longer uniform. Plots that are too small may prevent the accurate assessment of treatment effects. If the space available for an experiment is limiting, having more replications is usually more beneficial than having larger plots as long as your plot size allows accurate assessment of treatment effects.

EXPERIMENTAL DESIGNS

Completely Randomized Design

The completely randomized design is the simplest experimental design. In this design, treatments are replicated but not blocked, which means that the treatments are assigned to plots in a completely random manner (as in the left side of Fig. 6). This design is appropriate if the entire test area is homogeneous (uniform in every way that can influence the results). Unfortunately, it is rare that you can ever be confident of a test site's uniformity, so a completely randomized design is rarely used in field tests. The completely randomized design is used more commonly in greenhouse tests, though blocking is often useful even in the more controlled environment of a greenhouse.

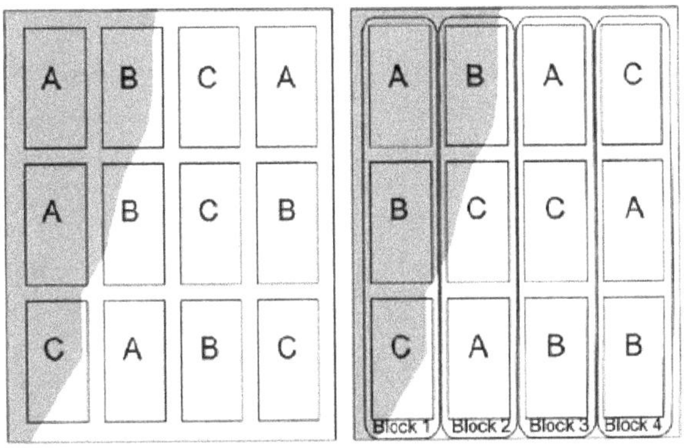

Fig. 6. The shaded area represents an area of the field that is different from the unshaded area. Treatments (A, B, and C) are replicated but not blocked in the field on the left. On the right, treatments are replicated and blocked; each block contains one plot of each treatment.

Randomized Complete Block Design

The randomized complete block design is the most commonly used design in agricultural field research. In this design, treatments are both replicated and blocked, which means that plots are arranged into blocks and then treatments are assigned to plots within a block in a random manner (as in the right side ofFig. 6). This design is most effective if you can identify the patterns of non-uniformity in a field such as changing soil types, drainage patterns, fertility gradients, direction of insect migration into a field, etc. If you cannot identify the potential sources of variation, you should still use this design for field research but make your blocks as square as possible. This usually will keep plots within a block as uniform as possible even if you cannot predict the variation among plots.

Blocking refers to physically grouping treatments together in an experiment to minimize unexplained variation in the data you collect (referred to as experimental error). This allows the statistical analysis to identify treatment differences that would otherwise be obscured by too much unexplained variation in the experiment. Variation in an experiment can be divided into two types: variation for which you can account in the statistical analysis and variation that is unexplained. The goal in blocking is to allow you to measure the variation among blocks and then remove that variation from the statistical comparison of treatment means. If you can anticipate causes of variation, you can block the treatments to minimize variation within each block and remove some variation from the statistical analysis. The mathematics of how blocking allows you to reduce unexplained variation is beyond the scope of this bulletin.

In the most common experimental designs, a block will contain one plot of each treatment in the experiment. If an experiment has five treatments, then each block will contain five plots, with each plot receiving a different treatment. When a block contains one plot of each treatment, then each block represents one replication of each treatment. For this reason, blocks are frequently referred to as "replications" or "reps," but the concept of blocking should not be confused with the concept of replication; replication and blocking serve different purposes. In agricultural research, field plots are almost always blocked even when no obvious differences are present in the field. It is much better to block when you did not really need to than not to block when you should have blocked.

Blocking is a very powerful tool that is most effective if you can anticipate sources of variation before you begin an experiment. For example, in a herbicide trial, one side of a field may have a history of more severe weed problems. If you just scattered your treatments randomly through the field, a lot of the variation in the data you collected could be due to the increased weed pressure on one side of the field. Such variation would make it difficult to determine how well each treatment worked. Because you know one side of the field will have more weeds, you can remove that source of variation from the statistical analysis by blocking and improve your chances of identifying differences among treatments.

The process of blocking follows a logical sequence. First, you determine that there is something (weeds, drainage, sun/shadow, water, soil type, etc.) that is not uniform throughout the experimental area (field, greenhouse, etc.) that may influence whatever you are measuring (yield, plant height, etc.). Then you can arrange your treatments into blocks so that the area within each block is as uniform as possible (see figure 6). Though the area within a block should be relatively uniform, there may be large differences among the blocks, but that is what makes blocking effective. Your goal is to maximize the differences among blocks while minimizing the differences within a block.

The shape of the blocks is not important as long as the plots within a block are as uniform as possible. Ideally, the only differences among plots within a block should be due to the treatments. Blocks in field experiments are usually square or rectangular, but they may be any shape. Blocks in the same experiment do not have to be the same shape; the shape of individual blocks will be determined by variation in the field that you are trying to minimize. If you are not sure what shape your blocks should be, square or nearly square blocks are usually a safe choice.

Blocks may be arranged through the field in many ways. If the field is wide enough, an easy way to arrange blocks is to place them side-by-side all the way down the field (see figure 7). But blocks do not have to be contiguous and may be scattered through the field in any way that is convenient for you.

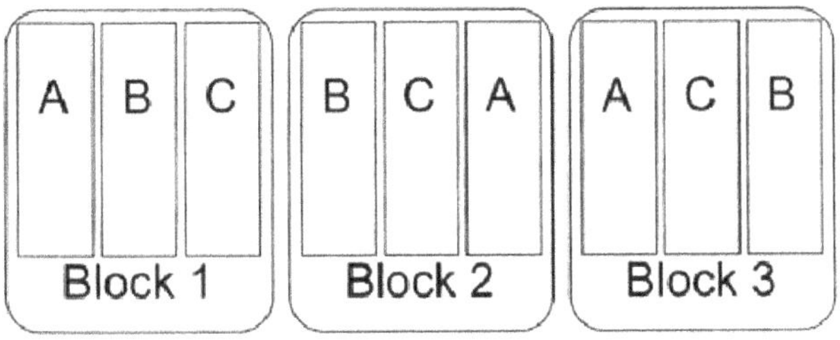

Figure 7. An easy way to arrange blocks is to put them side by side across the field. Letters represent different treatments.

FACTORIAL ARRANGEMENT OF TREATMENTS

A factorial arrangement of treatments is not an experimental design, though you will often hear it referred to as a factorial design or a factorial experiment. A factorial arrangement of treatments means that the experiment is testing two or more factors at the same time, and that the experiment includes all combinations of all factors. The term "factor" is used to describe a group of treatments that have something in common. Fungicides, sources of nitrogen, or corn hybrids could be considered factors in an experiment. Factors may be defined broadly or narrowly in different experiments. All herbicides may be grouped as a factor in one experiment, but pre-plant and post-plant herbicides may be treated as separate factors in another experiment. A single-factor experiment tests one factor at a time; a two-factor experiment tests two factors at once.

Most simple on-farm experiments are single-factor experiments (in a Completely Randomized or Randomized Complete Block design) and compare things such as crop varieties or herbicides, but it is sometimes useful to test two or more factors at once. For example, a two-factor experiment would allow you to compare the yields five corn hybrids at three planting dates. This accomplishes three things at once:

1. It allows you to compare the corn hybrids with each other.
2. It allows you to evaluate the effect of planting date.
3. It allows you to determine if varying the planting date changes the relative performance of the hybrids (e.g. one hybrid may only perform well if planted early).

The first two could be done in separate single-factor experiments, but the third can only be achieved by having both factors in a single experiment. This becomes especially important if one factor can have a significant influence on the effect of the other factor. For example, you might test soybean varieties as one factor and nematicides as another factor. If a few varieties have good nematode resistance but others do not, they may appear equally good when effective nematicides are used but varieties with resistance would appear much better when nematicides are not used. In cases like this, the effect of one factor (variety) is strongly influenced by the other factor (nematicide). When one factor influences the effect of the other factor, there is said to be a significant interaction between the two factors. It can be very important to know if there is an interaction between factors, because if there is an interaction, you can make predictions or recommendations based on the results of single-factor experiments ONLY when all other factors are at the same levels they were at in the experiment. If you change some factor not included in the experiment, the results from your single-factor experiment may no longer be valid.

With a factorial arrangement of treatments, all values (or levels) of each factor must be paired with all levels of the other factors. If you have two nematicides and five soybean varieties, then your treatment list must include each variety with each nematicide for a total of 10 treatments. This would be referred to as a "two by five factorial" to denote how many factors were present in the experiment and how many levels of each factor were used. The number of treatments increases quickly when you add more levels for a factor (if you used three nematicides instead of two, you would have 15 treatments instead of 10), so choose your levels carefully or the experiment can get too large to manage.

A factorial arrangement of treatments can be a very powerful tool, but because the number of treatments can get very large it is best used when some reason exists to believe that the factors may influence each other and have a significant interaction. If there is no suspicion that the factors may influence each other, it is frequently easier and more thorough to test the factors in separate experiments. A factorial arrangement of treatments can be used with a completely randomized experimental design or a randomized complete block design.

SPLIT-PLOT EXPERIMENTAL DESIGN

A split-plot experimental design is a special design that is sometimes used with factorial arrangements of treatments. This design usually is used when an experiment has at least two factors and some constraint prevents you from randomizing the treatments into a randomized complete block design. Such a constraint may be based on equipment limitations or on biological considerations. For example, the equipment you have may make it difficult to put out a soil fumigant in randomized complete blocks, but you may be able to put out the fumigant so that all treatments within a block that get the fumigant will be clustered together rather than scattered throughout the block. You can use a split-plot experimental design to work around this limitation as long as you are able to randomize the other factors. There are other situations when this design is appropriate, but a constraint on randomization is the most likely to occur.

Suppose you want to test the effect of five fungicides to control Cylindrocladium Black Rot on two varieties of peanut. In this test, you would have a 2x5 factorial arrangement of treatments: The two factors would be varieties (2 levels of this factor) and fungicides (5 levels of this factor). Because a factorial arrangement of treatments is not an experimental design, you still have to select an experimental design that best meets your needs. If you are able to randomize varieties and fungicides within a block, then you should pick a randomized complete block design. If there is some reason why you cannot completely randomize the treatments within each block, then you may be able to use a split-plot design to work around that limitation. For example, you may have a six-row planter but only enough space in the field to put out four-row plots. To resolve this dilemma, you could plant all of the plots that have the same peanut variety together within a block and then randomize the five fungicide treatments within each peanut variety.

In split-plot designs, the terms "whole plots" and "sub-plots" refer to the plots into which the factors are randomized. As the names imply, whole plots are subdivided into subplots. In figure 4, a whole plot would be the areas designated with A or B, and the subplots, the subdivisions within the whole plots, are designated 1, 2, 3, 4, or 5. In this example, A and B could represent two varieties (two levels of one factor) and the numbers could represent different fungicides (five levels of a second factor). Each whole plot serves as a block for the subplot treatments.

To assign treatments in a split-plot design, start by identifying where each block will be. Then randomize the whole plot treatments within each block. The whole plot treatments will be the treatment that you are unable to randomize into a randomized complete block design. The subplot treatments can then be randomized within each whole plot treatment (see figure 4).

EFFECT OF FUNGICIDE APPLICATIONS ON GRAIN SORGHUM (SORGHUM BICOLOR L.)

Fungicide Application

At the Wharton County location, fungicides were applied with a Lee® Spider sprayer equipped with one 8003 XR flat fan nozzle (TeeJet Spraying Systems Co., Wheaton, IL 60188) per row, while at Alexandria, fungicides were applied with a CO_2-propellant backpack sprayer equipped with one 8001 flat fan spray nozzle (TeeJet Spraying Systems Co., Wheaton, IL 60188) per row. At all locations, fungicides were applied in 140 L of water/ha at a pressure of 315 kPa at the Wharton County location or 504 kPa at Alexandria. Fungicides were applied at 25% bloom at each location which was 84 (May 31), 80 (June 23), and 83 (June 23) days after planting (DAP) in 2013, 2014, and 2015, respectively.

Leaf Chlorophyll Concentration

Grain sorghum leaf chlorophyll concentrations were determined using a SPAD 502 chlorophyll meter (Konica Minolta; Ramsey, NJ 07446). Readings were taken on the flag leaf between the mid-vein and leaf margin and reported values are the average of readings collected from ten plants in the center two rows of each plot at solar noon. Measurements were taken on a weekly basis beginning the day before fungicide application until black layer was reached.

Leaf Temperature

Leaf temperature readings were taken using a Raytek ST Pro temperature gun (Raytek Corp.; Santa Cruz, CA 95061). Readings were taken at the same location and time as leaf chlorophyll measurements on ten plants and reported values are the average of the readings.

Seed N and Protein, Stalk Lodging, and Grain Mold Ratings

Seed N and protein were determined by the high temperature combustion process. Stalk lodging data was collected the day prior to harvest and was calculated based on the total number of plants lodged per plot divided by the total number of plants per plot. Grain mold was determined on a scale of 1 to 5 with 1 = seed bright with no mold or discoloration and 5 = seed covered entirely with mold and was deteriorated and looked dead.

Data Analysis

The treatment design was a factorial arrangement using a randomized complete block design with fungicides and grain sorghum hybrids as factors. An analysis of variance was performed using the ANOVA procedure for SAS (SAS Institute, 1998, SAS user's guide, SAS Institute, Cary, NC) to evaluate the significance of fungicide and grain sorghum hybrid on overall plant health which included leaf chlorophyll concentration and temperature, plant lodging, grain mold, grain test weight and yield, and percent nitrogen and protein in the grain. The Fishers Protected LSD at the 0.05 level of probability was used for separation of mean differences.

Leaf Chlorophyll Content

The effect of fungicide was only noted in 2015 at 13 weeks after planting when the untreated and flutriafol treatments produced the highest readings and pyraclostrobinthe lowest. Also, hybrid response was noted in 2013 at 13 weeks after planting when DKS 53-67 resulted in higher readings than DKS 54-00 and also in 2013 and 2014 at 15 weeks after planting when different results were noted. In 2013, DKS 53-67 resulted in the highest reading while in 2014 the opposite was found. Researchers noted no difference in leaf chlorophyll readings with nine different fungicides on soybean.

Leaf Temperature Readings

Only in 2015 at week 11 was a response noted with fungicides while hybrid differences were noted in 2013 and 2014. In 2015 at week 11, leaf temperature was the greatest with pyraclostrobin while azoxystrobin produced the lowest temperature. Hybrid differences in leaf temperature were noted in 2013 and 2014 but not 2015. Results were inconsistent in both years as in some instances DKS 53-67 produced higher leaf temperatures and in other instances DKS 54-00 produced higher temperatures.

Test Weight

Only a hybrid response was noted in 2013 and 2014 with DKS 53-67 producing higher test weight than DKS 54-00 in both years.

Yield

A response to fungicides was noted only in 2015. Flutriafol produced the highest yield and azoxystrobin and pyraclostrobin the lowest. However, no fungicide treatment was different from the untreated check. In 2013 and 2014, DKS 53-67 outyielded DKS 54-00.

Researchers noted, in a 2-year study on soybean, that fungicides had no effect on yield. Grichar noted that the use of fungicides on soybean along the upper Texas Gulf Coast under little or no disease pressure resulted in few increases in yield and subsequent increases in net returns. In several instances, the use of fungicides resulted in a decrease in net returns, especially in a year with below normal rainfall. However, researchers reported that, in corn, generally the mean yield was higher in plots treated with fungicides than in the nontreated plots.

Protein Content

Only a hybrid was noted with DKS 53-67 having a higher protein content than DKS 54-00 in both 2013 and 2015 with no differences noted in 2014. Researchers noted no differences in protein content among fungicide treatments in soybean.

Seed N Content

As was seen with protein content, only a hybrid response was noted. In all three years, DKS 53-67 had a higher N content than DKS 54-00.

Grain Mold

In 2015, the untreated check resulted in a higher grain mold rating than any of the fungicide treatments while in 2014 a hybrid difference was noted as DKS 54-00 had a higher rating than DKS 53-67.

Lodging

The results from the various variables tested in this study fail to show a consistent overall improvement in plant health and other advantages to using a fungicide under low disease pressure. Other studies in various crops have also failed to indicate that the application of a fungicide would improve overall plant health and result in an increase in yield.

The prophylactic use of fungicides may confer risks beyond economic losses. QoI or strobilurin fungicides are classified by the Fungicide Resistance Action Committee (FRAC) as high-risk for resistance development. Over 40 pathogens have been reported as being resistant to QoI fungicides worldwide.

Recently, isolates of *Cercospora sojina* Hara on soybean were confirmed as resistant to strobilurin fungicides even though these fungicides have only been widely used in soybean for a few years. Researchers concluded that under the low disease incidence noted in Iowa a low probability exists that the use of a fungicide on soybean would increase yield by mechanisms other than disease control.

Researchers reported that foliar applications of azoxystrobin may be useful for the management of some foliar soybean diseases, but azoxystrobin may increase the percent of *Phomopsis* spp. seed infection. It was felt that azoxystrobin may interfere with the plant's natural defense mechanism to *Phomopsis* spp., or it may protect the plant from other diseases, thus extending the life of the plant so that *Phomopsis* spp. has more time to move from the pod into the seed. Researchers reported no long-term negative impacts on the soil system or strawberry (*Fragaria* x ananassa) production as a consequence of strobilurin usage. In one of two years, they reported a yield boost on strawberries as a result of fungicide usage.

Researchers stated that unless a corn crop is at risk of developing fungal disease, farmers would be smart to skip fungicide treatments that promise increased yields. They reported that fungicides used in fields where conditions were optimal for fungal diseases improved yields and paid for themselves. Some studies in wheat *(Triticum aestivum)* have indicated that QoI and DMI fungicides may delay senescence causing a "greening effect" by reducing oxygen in the leaves thereby protecting plants from toxic reactive oxygen species.

Others have hypothesized that the greening effect in wheat is a result of the inhibition of ethylene production by the QoI fungicides. Other research has found no effect of the QoI or DMI fungicides on wheat senescence, biomass, or yield.

In conclusion, in the absence of any disease pressure, the application of a fungicide may not improve grain sorghum plant health or increase yield and may ultimately reduce net returns because of the added cost of a fungicide and application. However, if conditions are conducive to disease development, the application of a fungicide may prove beneficial.

5

ROLE OF FUNGICIDES IN AGRICULTURE, AND HORTICULTURE

Fungicides are pesticides that kill or prevent the growth of fungi and spores. They can be used to control fungi that damage plants, such as rusts, mildews, and blights. They might be used to control mold and mildew in other settings. Fungicides work in a variety of methods, but most of them damage fungal cell membranes or interfere with energy production within fungal cells.

- What is fungicide in agriculture
- What are the different types of fungicides
- How do you apply a fungicide to plants
- Are fungicides harmful
- Is there a natural fungicide
- Fungicides in plant disease control
- Do fungicides kill bacteria

Fungicides, herbicides, and insecticides are all pesticides used in crop protection. A fungicide is a specific type of pesticide that controls fungal disease by specifically inhibiting and killing the fungus causing the disease. Let us get into the full details of the role of fungicides in agriculture crops and garden plants.

A STEP BY STEP GUIDE TO THE ROLE OF FUNGICIDES IN AGRICULTURE CROPS

Not all diseases caused by fungi can be adequately controlled by all fungicides. These contain the vascular diseases Fusarium and Verticillium wilt. Diseases caused by other types of organisms, disorders caused by abiotic factors, and insect damage are not controlled by these fungicides. Therefore, it is essential to first determine the cause of symptoms before applying a fungicide. The development of botanical fungicides can help decrease the negative impacts of synthetic agents, such as residues, resistance, and environmental pollution. In this respect, botanical fungicides in agriculture can be effective, selective, biodegradable, and less toxic to the environment. Fungicide is also called antimycotic, any toxic substance used to kill or inhibit the growth of fungi. Most agricultural fungicides and horticultural fungicides are applied as sprays or dust. Seed fungicides are applied as a protective covering before seed germination.

Fig. 1

WHY ARE FUNGICIDES NEEDED AND WHY THE ROLE OF THEM IS IMPORTANT IN AGRICULTURE

Diseases are occurrence on plants, and having a significant economic impact on crop yield and quality, managing some diseases is an essential component of production for most crops. Broadly, there are three reasons fungicides are used:

a. To control disease during the establishment and development of a crop.
b. To increase the productivity of a crop and reduce blemishes. Diseased food crops can produce less because their leaves, which are required for photosynthesis, are affected by the disease. Blemishes can affect the edible part of the crop, in the case of ornamentals, their attractiveness, which both can affect the market value of the crop.
c. To improve the storage life and quality of harvested plants. Some of the greatest disease losses occur post-harvest time. Fungi spoil (render unusable) stored fruits, vegetables, tubers, and seeds. A few which infect grains make toxins (mycotoxins) capable of causing severe illness or even death in humans and animals when consumed. Fungicides have been used to reduce mycotoxin contamination in wheat affected by Fusarium head blight, but most fungicides developed so far have not been effective to be useful for managing mycotoxins associated with other diseases.

THE ROLE OF FUNGICIDES IN AGRICULTURE

Fungicides are used to control bacterial and fungal diseases of fruit, vegetable, nut, and field crops. Some of the diseases that are controlled by using fungicides are mildew, leaf spots, blights, and apple scab. The fungicides are the last line of defense in the armoury of integrated disease management (IDM) system. They do not create crop yield but protect an inherent yield potential that the grower may realize in the absence of disease. Securing effective disease control from fungicide applications is dependent upon the disease pressure and the effectiveness of the fungicide to control the disease. Globally, the same fungicide active ingredients are mainly used against a similar range of fungal pathogens.

SOME TIPS FOR USING FUNGICIDES

Keep these tips in mind when using fungicides;

- Fungal disease in plants can be misdiagnosed easily and check with your local county extension office for help identifying plant disease. They may be able to recommend a treatment strategy for your lawn or garden.
- Plant diseases are transmitted when leaves are wet. Ground-level watering and good air circulation can be used to keep plant leaves dry.
- Pruning shears and tools can carry plant diseases from one plant to another. Learn about sanitation to prevent spreading fungal pathogens yourself.
- Although they can slow or stop the development of new symptoms, several fungicides are designed only to prevent disease. These are not highly effective after symptoms have developed.
- Apply fungicides before the development of disease and many fungicides have systemic action they will not completely eradicate diseases after they have started. Most systemic fungicides move less than an inch toward the tip of the plant or just move from the upper to the lower side of the leaf.
- Apply fungicides before a rain if possible and water is necessary for most fungal spores to infect foliage and for the splash dispersal of spores. So, apply fungicides before a rain if it appears that the fungicide will have a chance to dry before the rain. Most fungicides have a good sticker and persist through rains pretty well.
- The timing of fungicide applications is important than nozzle type and spray pressure.
- Use copper products for bacterial diseases. For the most part, copper products are effective against bacterial diseases than they are against fungal diseases.
- Some diseases need specialized fungicides. Diseases, such as downy mildew and Phytophthora blight could require specialized fungicides. It can be wasteful to

apply specialized fungicides all season long for diseases that are not a threat. For example, downy mildew of cucurbits does not arrive in Indiana until late in the season.

ROLE OF FUNGICIDES IN DISEASE MANAGEMENT OF AGRICULTURE

Plant diseases are best managed by integrating several control practices that can include; crop rotation, selection of disease-tolerant or disease-resistant crop cultivars (cultivars genetically less susceptible than other cultivars), planting time, level of fertilization, micro-climate modification, sanitation, and application of fungicides. Fungicides are a vital part of disease management as;

4. Hey control many diseases satisfactorily,
5. Cultural practices do not provide adequate disease control,
6. Resistant cultivars are not available or not accepted in the marketplace for many diseases, and
7. Certain high-value crops have an extremely low tolerance for disease symptoms.

CLASSIFICATION OF FUNGICIDES

Fungicides fall into many categories that overlap. Some of the products stick to the plant on contact, providing a layer of protection against fungi; others are systemic and then move through the entire plant, destroying the fungi from inside the plant.

The second class of fungicides is split into preventative and curative products. Preventative fungicides protect the plant from fungal diseases, while curative fungicides only work to treat plants already suffering from a disease.

A single fungicidal product can belong to both categories, thus a contact fungicide prevents fungal spores from germinating on a plant. Conversely, a systemic fungicide can be preventative or it might be curative. What this means is that it is important to be clear on how the product works when you're applying a fungicide. A preventative fungicide is ineffective on plants suffering from a fungal infection, and a contact fungicide could wash off in rain and have to be regularly reapplied.

Fungicides differ in how they affect the fungal cells. Some target a particular procedure or range of processes within the cells, or they may damage the cell membrane. When the same fungicide is applied regularly to plants, fungi develop resistance to its control system and the product loses effectiveness. To help stop resistance from developing in the garden, use a range of products that employ different methods to control fungi.

Some chemical-based fungicides are Mancozeb, Tricyclazole, Carbendazim, Hexaconazole, Metalaxyl, Benomyl, Difenoconazole, Propiconazole, Kitazin, Tebuconazole, Copper oxychloride, Copper hydroxide, Tridemorph, and Propineb

The other category is an organic type of fungicide. They are Safin, Sporrin, Blastin, and Bio-Vitrioll.

Fungicides are pesticides that prevent, kill, mitigate, or inhibit the growth of fungi on plants, but they are not effective against bacteria, nematodes, and viral diseases. Fungicides can be mainly classified based on:

Mobility in the plant

Contact fungicides are not absorbed by the plant and then stick to plant surfaces. They give a protective barrier that prevents the fungus from entering and damaging plant tissues. Systemic products also called penetrants, are absorbed by the plant and can move from the site of application to other parts of the plant. Movement in the plant changes by fungicide form moving to old tissues and new tissues, new growth, moving from the top to the bottom of the leaf surface.

Preventive vs. curative

These preventive fungicides work by preventing the fungus from getting into the plant. The preventive fungicides come into direct contact with the fungus, and they have to be re-applied to new plant tissues. Though, curative fungicides affect the fungus after infection. This means they can stop the disease after the infection has started and after the first symptoms are observed. Fungicides that can move in the plant can be preventative and curative.

Mode of action

This mainly refers to how the fungicide affects the fungus. Fungicides work by damaging the cell membrane of the fungus, inhibiting an important method that the fungi, pinpointing a single process, or multiple processes in the fungus. It's important to incorporate different modes of action by mixture or by alternating products to keep effectiveness and prevent fungicide resistance.

APPLICATION METHODS OF FUNGICIDES

Fungicides are applied as a dust, granules, gas, and, commonly, liquid. They are applied to;

a. Seed, bulbs, roots of transplants, and other propagative organs. These treatments are done by the seed company. Some treatments require to be done by the grower on-site at the time of planting. The main goal is to kill pathogens that are on the planting material or to protect the young plant from pathogens in the soil.
b. Soil either in-furrow at planting, after planting as a soil drench (including through drip irrigation), and as a directed spray around the base of the plant.
c. Foliage and other aboveground parts of plants using a sprayer.

d. Inside of trees by trunk injection.
e. Harvested produce, as a dip or spray in the packinghouse.

Fungicides are used as a formulated product consisting of an active ingredient plus inert ingredients that develop the performance of the product. Fungicides are mixed with water then applied by spraying. Application equipment ranges from small hand-held and back-pack sprayers to large spray units carried by tractors. A few fungicides are mainly applied as dust. Fungicides can be applied in greenhouses as smoke, mist, fog, or aerosol. Coverage of all parts of the plant susceptible to the disease is critical as very few fungicides can move adequately throughout a plant. Advancements are continually being made to nozzles and sprayers to improve coverage.

NATURAL FUNGICIDES ROLE FOR GARDEN PLANTS OR AGRICULTURE

A range of naturally based fungicide products is available for your gardens. Copper, sulfur, and lime commonly used in powder or oil form are extremely effective at treating and preventing fungal infections. Some gardeners prefer to go even more natural and use products found around the home instead.

Oils

Oils are used to control insect pests, but by controlling these, you also manage the viruses and fungal infections they can cause. Oils are effective in managing powdery mildew disease on a variety of plants. Cooking oils found in homes, such as canola and vegetable oil, can kill disease-carrying insects. This, in turn, helps to prevent fungal infections. Oil sprays can burn the leaves of sensitive plants, thus use no more than 1 cup per gallon of water when mixing for spray use.

Bicarbonates

Sodium bicarbonate or baking soda is natural fungicides that has been used for decades, but avoid using baking soda with horticultural oils because the sodium component can build up and become toxic to plants. Baking soda alone is effective against a limited number of fungal diseases, however. The use of ammonium or potassium bicarbonate, along with neem oil, provides a better, and more effective, treatment because it provides nitrogen.

Companion Planting

One of the natural fungicides available to you is other plants. Companion planting can help fight fungal disease and then improve the soil as well. Borage, for example, benefits nearly any plant growing nearby by increasing its resistance to pests and disease. It is mainly helpful to strawberry plants and tomatoes. Chives, which improve the growth and flavor of vegetable crops, could help prevent scab in apple trees and black spot in roses. Do not plant chives with beans or peas. Chives take about 3 years to begin effectively preventing disease, so you may want to use another natural fungicide until the chive plants are well-established. Garlic plants accumulate sulfur in the soil, which is a natural fungicide that plants will absorb into their pores when it is planted nearby.

Compost Teas

Compost teas have beneficial microbes and nutrients that can act as mild fungicides, and they provide control of diseases carried by insects. To make compost tea, add 2 inches of finished compost into the bottom of a bucket. Mix in an ounce of unsulfured molasses to feed the beneficial microorganisms in the compost and fill the bucket to 6 inches from the top of the bucket with water. Let the mixture steep outdoors in the sunlight for 2 to 3 days and then strain the compost from the tea with cheesecloth. The tea must smell sweet and earthy. If it smells rancid or sour, don't use it on plants. Pour it back into the compost pile and begin again. Spray the tea onto your plants and the soil about every 2 to 3 weeks.

6

FUNGICIDE STEWARDSHIP

FUNGICIDE RESISTANCE

Fungicides are important tools for preventing and managing plant disease. Unlike insecticides and some herbicides which kill established insects or weeds, fungicides are most commonly applied to protect healthy plants before infection occurs. Resistance to fungicides is one of many possible causes of poor disease control (see Is Resistance to Blame?). Resistance refers to a situation where a given fungicide once controlled a particular fungal population but, after one or more applications, that fungicide no longer controls that population.

MECHANISMS OF FUNGICIDE RESISTANCE

There are several ways that populations of fungi can become resistant to fungicides, these include:

1. **Altered target site.** A fungicide has a specific target site where it acts to disrupt a particular biochemical process or function. If this target site is somewhat altered, the fungicide no longer binds to the site of action and is unable to exert its toxic effect. This is the most common mechanism that fungi use to become resistant. See the example below.
    i. **Enzyme** A is necessary for carbohydrate metabolism in fungi. The target site is the point where the enzyme is used.

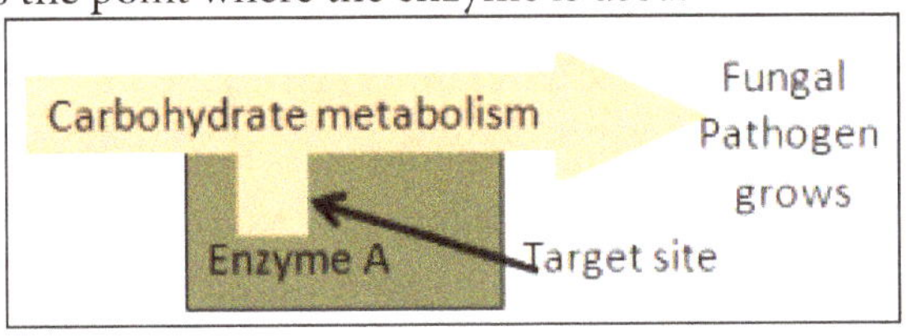

Fig. 1

ii. Fungicide A interferes with carbohydrate metabolism by filling the target site of Enzyme A.

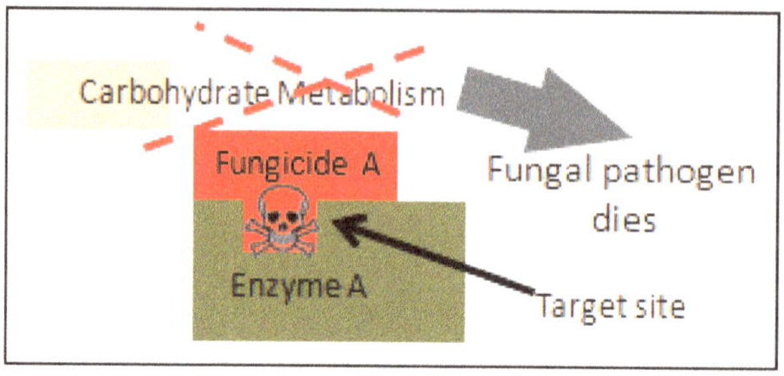

Fig. 2

iii. In resistant fungi, a part of the site is altered preventing the fungicide from fitting, but allowing carbohydrate metabolism to occur.

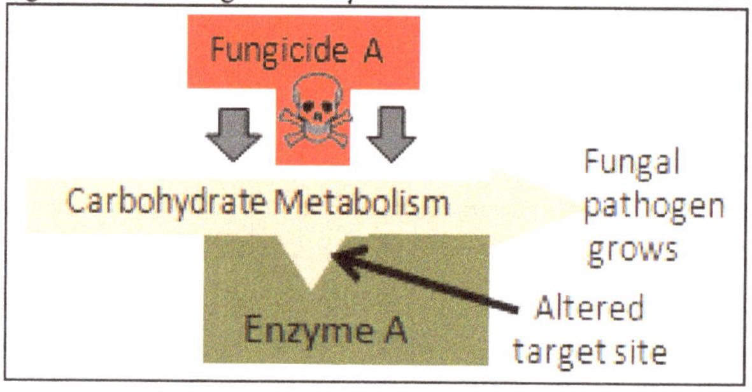

Fig. 3

2. **Detoxification or metabolism**. Metabolism within the fungal cell is one mechanism a disease pathogen uses to detoxify a foreign compound such as a fungicide. A fungus with the ability to quickly degrade a fungicide can potentially inactivate it before it can reach its site of action.

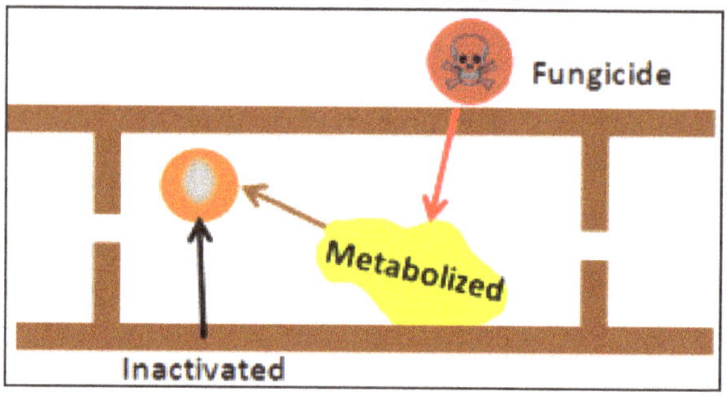

Fig. 6.4 Fungal cell

FUNGICIDE RESISTANCE MANAGEMENT

Fungicides are important tools for managing diseases in many crops. Unlike insecticides and some herbicides which kill established insects or weeds, fungicides are most commonly applied to protect healthy plants from infection by fungal plant pathogens. To be effective, fungicides must be applied before infections become established and in a sufficient spray volume to achieve thorough coverage of the plant or treated area. Protection from fungicides is temporary because they are subject to weathering and breakdown over time. They also must be reapplied to protect new growth when disease threatens. Poor disease control with fungicides can result from several causes including insufficient application rate, inherently low effectiveness of the fungicide on the target pathogen, improper timing or application method, and excessive rainfall.

Resistance (lack of sensitivity) to fungicides in fungal pathogens is another cause of poor disease control. The development of fungicide resistance is influenced by complex interactions of factors such as the mode of action of the fungicide (how the active ingredient inhibits the fungus), the biology of the pathogen, fungicide use pattern, and the cropping system. Understanding the biology of fungicide resistance, how it develops, and how it can be managed is crucial for insuring sustainable disease control with fungicides.

The problem of fungicide resistance became apparent following the registration and widespread use of the systemic fungicide (see fungicide mobility below) benomyl (Benlate) in the early 1970's. Prior to the registration of benomyl, growers routinely applied a protectant fungicide such as maneb, mancozeb, or copper to control diseases without experiencing resistance problems. A distinct advantage of benomyl over the protectant fungicides was its systemic activity.

In addition to protecting plants from infection, systemic activity conferred rainfastness and provided disease control when applied after the early stages of infection. Superior disease control was often achieved with benomyl compared to the protective dithiocarbamates.

However, benomyl differed from the dithiocarbamates in its site-specific mode of action which was readily overcome by several fungal pathogens. Resistance problems appeared a few years after benomyl was introduced where the fungicide was used intensively. Sudden control failures occurred with diseases such as powdery mildew, peanut leaf spot, and apple scab.

Many of the fungicides developed and registered since the introduction of benomyl also are systemic, have a site-specific mode of action, and are at increased risk for resistance problems.

Fungicide resistance is now a widespread problem in global agriculture. Fungicide resistance problems in the field have been documented for nearly 200 diseases (crop – pathogen combinations), and within about half of the known fungicide groups.

Many more cases of resistance are suspected but have not been documented. While resistance risks with many of fungicides may not be as great as with benomyl, strategies to manage the resistance risk have been developed and implemented to avoid unexpected control failures and sustain the usefulness of new products. As a result of resistance management strategies, fungicides within all mode of action groups remain useful disease management tools in at least some cropping systems.

Since this fact sheet was first written, many new fungicides have been registered, and mode of action groups and specific resistance management strategies are now specified on fungicide labels. The listing of fungicides by mode of action group here is useful for identifying appropriate fungicides for use in tank mixtures and application schedules as part of the recommended resistance management programs.

Fungicide Mobility

Understanding the mobility of fungicides on and in treated plants, and how various fungicides are classified based on mobility is important when making decisions pertaining to the selection of the best of fungicide for a particular disease and its optimal application timing. Fungicides can be classified into two basic mobility groups: protectant or penetrant. Regardless of its mobility characteristics, no fungicide will be highly effective after the development of disease symptoms and pathogen reproduction (spore production). Fungicides can slow or stop the development of new symptoms if applied in a timely fashion, but fungicides will not cure existing disease symptoms. Therefore, understanding fungicide mobility, fungicide mode of action, and the biology of the target pathogen are important so that fungicide applications are made before the disease becomes established and more difficult to control.

Protectant fungicides are active on the plant surfaces where they remain after application. There is no movement of the fungicide into the plant. Because they remain on the plant surface, protectant fungicides loose activity after being washed off the plant and must be re-applied to new growth that develops after application. Protectant fungicides typically prevent spore germination, therefore they must be applied prior to infection and have no effect once the fungus grows into the plant resulting in infection.

Penetrant fungicides are absorbed into plants following application. Because these fungicides are absorbed into plants, they are generally considered systemic fungicides. However, penetrant fungicides have different degrees of systemic movement once inside the plant. Some fungicides are 'locally systemic', only moving a short distance such as through a few layers of plant cells. Fungicides that move from one side of a leaf to other have 'translaminar' movement. Translaminar and locally systemic fungicides are not transported throughout the plant. Highly mobile fungicides are either 'xylem-mobile' or 'true systemics'. Xylem-mobile fungicides move upward in plants and outward to the periphery of leaves with water through the xylem, the water conducting tissue of the plant. True systemic fungicides move both upward through the xylem, and downward through the phloem, the food conducting tissue of the plant. Few if any fungicides are fully systemic. Unlike protectant fungicides, penetrant fungicides are rain fast within a few hours of application and may require less thorough application coverage to be effective. In addition, many penetrant fungicides inhibit fungal growth and sporulation and can be effective when applied after the early stages of infection.

Regardless of the level of systemic movement, penetrant fungicides have limited 'curative' ability. Generally they only stop or slow infections within the first 24 to 72 hour period following fungal penetration into the plant. Therefore, penetrant fungicides must be applied before or shortly after infection, and are ineffective on existing symptoms. Both protectant and penetrant fungicides provide good disease control when applied before infection and are best applied on a preventive schedule.

DEVELOPMENT OF FUNGICIDE RESISTANCE

Resistance is a genetic adjustment by a fungus that results in reduced sensitivity to a fungicide. Reduced sensitivity is thought to be a result of genetic mutations which occur at low frequencies (one in a million or less) or of naturally occurring sub-populations of resistant individuals. Individuals in a fungal population may consist of the mycelium (the body of a fungus), sclerotia (large survival structures), spores (small reproductive structures), or the nucleus of single cells capable of reproduction and spread.

The resistance trait may result from single gene or multiple gene mutations (see build-up of resistance below). Single-gene mutations that confer resistance to site-specific fungicides are more likely to develop than the simultaneous occurrence of mutations in multiple genes needed to confer resistance to multi-site inhibiting fungicides. Mechanisms of resistance differ depending on the mode of action, but include alteration of the target site, reduced fungicide uptake, active export of the fungicide outside fungal cells, and detoxification or breakdown of the fungicide.

The level of resistance to a fungicide can be measured in the laboratory by exposing a collection of members of a field population to the fungicide and measuring toxicity response. Toxicity responses are usually measured as inhibition of fungus growth, spore germination, or actual plant infection in cases where the fungus cannot be cultured. The effective concentration which inhibits growth, germination, or infection by 50% (EC50) is then calculated for each sampled individual much in the same way an LD_{50} (50% lethal dose) is calculated for assessing the acute toxicity of a pesticide to rats or mice. Where many members of a population are sampled and screened, a range of sensitivity (or resistance) to the fungicide is usually observed.

The frequency distribution of the sensitivity of individuals in the population is usually normal or bell-shaped, typical of many biological responses in nature (Figure 5). Where the fungicide is newly introduced or where the risk of resistance is low, the population is distributed over a sensitive range. However, a distribution consisting of two distinct sub-populations also may occur where a small sub-population of resistant strains is present along with a larger sub-population of sensitive strains (Figure 5).

BUILD-UP OF RESISTANCE

Resistance in a population becomes important when the frequency of resistant strains builds up to dominate the population. The build-up of resistant strains is caused by repeated use of the fungicide which exerts selection pressure on the population. The fungicide selectively inhibits sensitive strains, but allows the increase of resistant strains. This shift toward resistance occurs at different rates depending on the number of genes conferring resistance. When single gene mutations confer resistance, a rapid shift toward resistance may occur, leading to a population that is predominantly resistant and where control is abruptly lost (Figure 5A). When multiple genes are involved, the shift toward resistance progresses slowly, leading to a reduced sensitivity of the entire population (Figure 5B). The gradual shift with the multiple gene effect may result in reduced fungicide activity between sprays, but the risk of sudden and complete loss of control is low. It is difficult to clearly distinguish between sensitive and resistant sub-populations with field sampling during the early shifts towards reduced sensitivity because sensitivity responses overlap. Large numbers of individuals must be tested to identify the gradual type of resistance.\

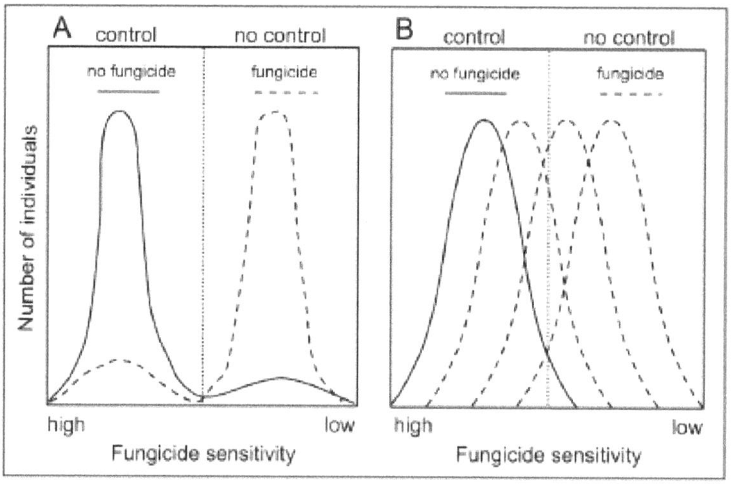

Fig. 5. Depiction of the possible ways fungicide resistance develops in population of a fungal pathogen.

A) Abrupt (qualitative) resistance development where an initially small, subpopulation of resistant strains is present before fungicide usage or develops as a result of a single gene mutation occurring at low frequency (solid line). Following selection pressure of fungicide use, the frequency of resistant individuals (broken line) becomes predominant and disease control is rapidly lost. B) Gradual (quantitative) resistance development arising from an accumulation of mutations in multiple genes that leads to reduced sensitivity. The initial population (solid line) is sensitive, but gradually shifts towards reduced sensitivity under the selection pressure of fungicide use (broken line).

ASSESSING RESISTANCE RISK

Many factors affect the development of resistance and its build-up in the field, which makes it difficult to predict the resistance risk for new fungicides. Despite resistance problems that have been identified following the introduction of some new fungicides, many examples can be cited where their use continues to be effective. Factors that must all be considered in assessing resistance risk include the properties of the fungicide, the biology of the pathogen, and the crop production system where the fungicide is used.

FUNGICIDE GROUPS AND MODE OF ACTION

Fungicides are grouped by similarities in chemical structure and mode of action (Table 1). Site-specific fungicides disrupt single metabolic processes or structural sites of the target fungus. These include cell division, sterol synthesis, or nucleic acid (DNA and or RNA) synthesis. The activity of site-specific fungicides may be reduced by single or multiple-gene mutations. The MBC (benzimidazole), PA (phenylamide), and QoI (strobilurin) groups are subject to single-gene resistance and carry a high risk of resistance problems. Other groups with site-specific modes of action include the Dicarboximides and DMIs (sterol demethylation inhibitors), but resistance to these fungicides appears to involve slower shifts toward insensitivity because of multiple-gene involvement. Many of the site-specific fungicides also have systemic mobility. However, systemic mobility is not necessary for resistance development. Resistance problems have developed in the dicarboximide group and with dodine which are protectant fungicides.

Multi-site fungicides interfere with many metabolic processes of the fungus and are usually protectant fungicides. Once taken up by fungal cells, multisite inhibitors act on processes such as general enzyme activity that disrupt numerous cell functions. Numerous mutations affecting many sites in the fungus would be necessary for resistance to develop. Typically these fungicides inhibit spore germination and must be applied before infection occurs. Multi-site fungicides form a chemical barrier between the plant and fungus. The risk of resistance to these fungicides is low.

FITNESS OF RESISTANT STRAINS

Fitness is the ability to compete and survive in nature. Strains of pathogens resistant to some fungicides compete equally well with sensitive strains and are still present after the fungicide in question is no longer in use. For example, strains of Cercospora arachidicola which cause early leaf spot of peanut are still established in the southeastern U.S. where benomyl resistance was a problem over 20 years ago. Therefore, fungicides with resistance problems cannot be successfully reintroduced into areas where resistant strains are highly fit. Fortunately, resistant strains are sometimes less fit than wild-type sensitive strains. This has been true for DMI resistance in powdery mildews and for dicarboximide resistance in Botrytis diseases. Unfit strains only compete well under the selection pressure of the fungicide. Thus, the resistance is at least partially reversible when the selection pressure of the fungicide is removed or minimized by using resistance management.

FUNGICIDE USE PATTERN

Frequent and exclusive usage of at-risk fungicides increases the risk of resistance problems. Selection pressure is increased where repeated applications are required for disease control as with many foliar diseases. Selection pressure and the risk of resistance are low for seed treatments and for many soilborne diseases which require only one or two applications per season. The method and rate of application may also impact resistance development. Poor disease control resulting from causes such as improper application timing or inadequate spray coverage may result in a need for a more intensive spray program and the exposure of more individuals to the fungicide. Using adequate rates in a manner that produces good disease control reduces the reproductive capacity of fungal pathogens, thus reducing selection pressure. Similarly, a preventive spray program is less risky than a rescue program because selection pressure is applied to fewer individuals. Finally, an increase in selection pressure results from an excessive number of applications where a real need is not justified.

PATHOGEN BIOLOGY

Fungal pathogens with high rates of reproduction are most prone to develop fungicide resistance. Because many individuals (usually spores) are produced by these fungi, more individuals are exposed to selection pressure and there is a greater probability of mutations that lead to reduced fungicide sensitivity. Foliar diseases produce thousands of spores on the surface of an individual leaf spot. Furthermore, these diseases typically have several reproductive cycles per season. Under selection pressure of a fungicide, resistant individuals may increase rapidly and dominate the population after several cycles of infection and reproduction.

Diseases with low reproduction rates generally complete only one life cycle per season. Soilborne pathogens produce fewer offspring per season than their foliar counterparts. Some soilborne diseases reproduce by forming seed-like survival structures called sclerotia. There may be fewer than a hundred sclerotia formed per plant. Where an at-risk fungicide is used for soilborne disease control, resistance development is likely to be slow because comparatively few individuals are exposed to selection pressure.

CROP PRODUCTION PRACTICES

Production practices that favor increased disease pressure also promote resistance development by increasing the number of individuals exposed to selection pressure. Pathogens reproduce at higher rates on susceptible varieties compared to resistant or partially resistant varieties. Selection pressure also may be reduced for resistant varieties because fewer applications should be needed for effective disease control.

Inadequate or excessive fertilization with nitrogen may increase disease incidence in some crops. For example, early blight of potato and tomato and dollar spot of turfgrass are favored by nitrogen deficiency. Alternatively, the severity of spring dead spot of bermudagrass and some foliar diseases of wheat is increased with intensive nitrogen fertilization. Excessive irrigation or frequent irrigation with small amounts of water increases the incidence of many diseases by promoting disease spread, extended periods of leaf wetness, and high soil moisture.

Continuous cropping and poor sanitation practices promote severe early-season disease development. Closed cropping systems such as greenhouses are particularly prone to resistance problems because plants are grown in crowded conditions that may favor severe disease development, rapid spread, and high selection pressure.

Permanently established plantings of perennial crops such as orchards, nurseries, and vineyards are also prone to resistance problems. Unlike annual crops where crop rotation can be practiced, many pathogens survive from year to year on plants and crop debris within permanent plantings resulting in a local pathogen population exposed to yearly selection pressures.

RESISTANCE MANAGEMENT STRATEGIES

Strategies for managing fungicide resistance are aimed at delaying its development. Therefore, a management strategy should be implemented before resistance becomes a problem. The only way to absolutely prevent resistance is to not use an at-risk fungicide. This is not a practical solution because many of the modern fungicides that are at risk for resistance problems provide highly effective, broad-spectrum disease control. By delaying resistance and keeping its level under control, resistance can be prevented from becoming economically important.

Because practical research in the area of fungicide resistance management has been limited, many of the strategies devised are based in the theory of expected responses of a pathogen population to selection pressure. For the most part, evaluations of the effectiveness of these strategies have not been based on research, but rather on observations made where the fungicides have been used commercially on a large scale.

Specific strategies for resistance management vary for the different fungicide groups, the target pathogen(s), and the crop. However, some strategies are generally effective. Resistance management should integrate cultural practices and optimum fungicide use patterns. The desired result is to minimize selection pressure through a reduction in time of exposure or the size of the population exposed to the at-risk fungicide. Probably the most important aspect of optimizing use patterns is the deployment of tank mixtures and alternating sprays of the at-risk fungicide with a fungicide from a different mode of action group.

The comparative merits of tank-mixing compared to alternating sprays have been debated. Some theorize that tank-mixing reduces selection pressure only when the partner fungicide is highly effective and good coverage is achieved. Alternating fungicides is thought to act by reducing the time of exposure. In practice, examples can be cited for the effectiveness of both approaches. Both practices are more effective when cultural practices are implemented to reduce disease pressure. The alternation of blocks of more than one spray is probably less effective in resistance management than the other use patterns.

For example, a block of four continuous sprays of the DMI fungicide tebuconazole is recommended at mid-season for peanut disease control. Despite the use of at least one application of a non-DMI fungicide before and after the 4-spray block, resistance to tebuconazole in both early and late leaf spot diseases became a widespread problem in less than 10 years.

The proper choice of a partner fungicide in a resistance management program is critical. Generally, good partner fungicides are multi-site inhibitors that have a low resistance risk (e.g. chlorothalonil, mancozeb, etc.) and are highly effective against the target pathogen. However, the use of an unrelated at-risk fungicide with no potential for cross-resistance problems also may be effective.

Numerous fungicides are now marketed as pre-mixtures of two or more fungicide groups. These are convenient for resistnace management, but users should be mindful of the fungicide groups in the pre-mix, so an at-risk group is not over-extended. Characteristics of important fungicide groups and their resistance risks are discussed below.

MBCs (FRAC Group 1; Mode of Action Sub-Group B1)

MBC (Methyl Benzimidazole Carbamate) fungicides are site-specific fungicides which interfere with cell division. They have systemic mobility and have activity on many pathogens except water molds (e.g. Pythium and Phytophthora) and darkly pigmented fungi (e.g. Alternaria). Research has demonstrated that benzimidazole resistant strains may be present at low frequencies in nature, even in the absence of fungicide exposure.

Under selection pressure, resistance development is abrupt and rapid (Figure 5A). Resistant strains cannot be controlled by increasing the application rate or by shortening the spray interval. Resistant strains are often fit and competitive in nature even without selection pressure. Therefore, some populations have remained resistant where benzimidazole use has been discontinued for 10 years. Resistance to benzimidazoles has been documented for over 70 diseases and cross-resistance exists within this fungicide group. Benzimidazole resistance has received less recent attention because the fungicide benomyl is no longer registered in the U.S.

However, resistance management remains important for thiophanate-methyl, the other widely used benzimidazole fungicide. In spite of the numerous resistance problems with benzimidazoles, there are also many examples where benzimidazoles have remained effective for over 30 years with judicial use.

Strobilurins (FRAC Group 11; Mode of Action Sub-Group C3)

Strobilurin fungicides, also know as quinone-outside inhibitor (QoI) fungicides, are synthetic analogues of a naturally occurring compound produced by a wood rotting fungus. Strobilurins inhibit respiration in fungal cells by targeting a protein (cytochrome bc-1) that is encoded by a gene in the mitochondria. The fungicides are broad-spectrum with activity against all the major types of fungal pathogens. Strobillurin fungicides penetrate plant leaves and move from one side of the leaf to the other. This translaminar mobility makes them rain-fast, but they lack true systemic movement in the plant compared to some other fungicide groups. Strobilurins act on a broad range of fungal processes including spore germination, fungal growth, and reproduction (sporulation).

Strobilurin fungicides have been registered on numerous crops because of their broad-spectrum activity and excellent human and environmental safety profiles. However, like the benzmidazoles, resistance developed shortly after their introduction in the late 1990's. Three different single-gene mutations have been identified that abruptly confer resistance (Figure 5A) that has been documented for more than 50 diseases. Resistance has been documented for diseases such as Septoria leaf spot on wheat; gummy stem blight, downy mildew, and powdery mildew on cucurbits, and frogeye leaf spot on soybeans. Resistant isolates are cross-resistant to all other strobilurin fungicides, but not to other mode of action groups including the closely related QiI (Group C4 or 21) fungicides. Several strobilurin fungicides are marketed in pre-mixtures with non-strobilurin fungicides for use on certain crops.

Dicarboximides (FRAC Group 2; Mode of Action Sub-Group E3)

Dicarboximides inhibit both spore germination and fungal growth. Resistance is thought to arise by mutations. The frequency of resistant individuals and their level of resistance increase gradually with prolonged selection pressure (Figure 5B). Resistance to dicarboximide fungicides has been identified for more than 15 diseases including brown rot of stone fruits, gray mold (Botrytis) on several crops, and important turf grass diseases. Dicarboximide resistant strains of some pathogens are less fit to survive than sensitive strains. Reduced exposure of resistant strains to dicarboximide fungicides results in a decrease in the frequency of resistant strains and possibly an overall shift of the population back toward sensitivity. Thus, it has been possible to reintroduce dicarboximides into problem situations where resistance management has been implemented.

DMIs (FRAC Group 3; Mode of Action Sub-Group G1)

Demethylation inhibitor (DMI) fungicides are site-specific fungicides that disrupt the synthesis of sterols. Sterols are compounds required for growth of many plant pathogenic fungi. DMIs are a large group of systemic fungicides that have a broad range of activity against many types of foliar and soilborne diseases except for those caused by the water molds. Typically, resistance develops gradually as a result of an accumulation of several mutations (polygenic resistance) and is at first difficult to detect (Figure 5B).

Resistant strains are thought to have reduced fitness; therefore, reduced selection pressure through the use resistance management strategies may partially shift the resistant populations back toward sensitivity. DMI resistance has been documented for over 30 diseases including apple scab, powdery mildews, gray mold, and brown rot of stone fruit.

Pre-mixtures of DMI fungicides with strobilurin or protectant fungicides are being marketed for many crops to improve the spectrum of diseases controlled and to comply with resistance management guidelines.

PAs (FRAC Group 4; Mode of Action Sub-Group A1)

PA (Phenylamide) fungicides are water soluble, highly systemic fungicides specifically used to control diseases caused by water molds. Such diseases include damping off and root and lower stem rots caused Pythium and Phytophthora, and foliar diseases such as late blight, downy mildew, and white rust. PA fungicides inhibit fungal growth by disrupting RNA synthesis.

Resistance problems with PAs, specifically metalaxyl, were observed shortly after their introduction where they were used exclusively and disease pressure was high. Resistance, now confirmed for over 30 diseases, is governed by one or two genes and a low frequency of resistant individuals may exist in wild populations prior to use of these fungicides. Resistance can increase rapidly through selection of the naturally occurring strains (Figure 5A).

Cross resistance occurs with other PA fungicides, but not with fungicides from other mode of action groups. Both resistant and sensitive strains survive in the absence of PA fungicide use and their levels tend to equilibrate over time. Resistance management is critical to limit the proportion of resistant strains in a population. The manufacturer of metalaxyl-M markets premixes with mancozeb, copper, and chlorothalonil for use against foliar pathogens. The marketing of pre-mixes of metalaxyl-M with non-related protectant fungicides ensures compliance with a resistance management strategy.

SDHIs (FRAC Group 7; Mode of Action Sub-Group C2)

SDHI (Succinate Dehydrogenase Inhibitor) fungicides inhibit respiration in fungal pathogens by blocking an important enzyme in mitochondrial respiration. SDHI fungicides are systemic and control a broad spectrum of disease. While the fungicide carboxin has been registered for over 40 years, several newer generation SDHI fungicides have been recently registered for use on a wide range of crops. SDHI fungicides are used as seed treatments as well as foliar applied treatments.

They have activity on a range of foliar diseases (rusts, powdery mildews, leaf spots) and soilborne diseases cause by Sclerotinia and Rhizoctonia. Resistance development, caused by several point mutations in the succinate dehydrogenase gene, has been documented for over 10 diseases such as gummy stem blight and powdery mildew on cucurbits, and grey mold on grapes and strawberries. Despite the few cases of documented resistance, the group has a moderate to high resistance risk and resistance management should be implemented for disease management programs that require multiple applications.

THE FRAC CODE

A proactive approach using diverse disease control tactics is the most effective way to avoid fungicide resistance. The primary objective of this preventive strategy is to reduce selection pressure by: 1) selecting and using fungicides correctly; and, 2) managing fields, farms, or sites wisely to reduce the inoculum of the pathogen and create conditions less favorable for the disease to occur. Doing what it takes to avoid fungicide resistance is a good alternative to paying the price for overcoming resistance once it occurs.

1. Select and use fungicides correctly:
   a. The best way to prevent resistance is to not use an at-risk fungicide. This is not a practical solution because many of the modern fungicides that are at-risk for resistance problems provide highly effective, broad-spectrum disease control. However, by reducing the number of single-site fungicide applications, you reduce the likelihood of resistance development. To minimize the impact of single-site fungicides, apply them in combinations as a tank mix or premix, 'twin pack' or 'combi- pack,' that provides multiple modes of action.
2. Tank-mixes and Pre-mixes are combinations of two or more fungicides applied as a single mixture. Tank-mixing allows for adjusting of the ratio of fungicides to fit local conditions, while premixes are formulated by the manufacturer. The combination, if composed of a high-risk (single-site) fungicide with a low-risk (multisite) fungicide will contribute to resistance avoidance. The different fungicides in the mixture must be active against the target fungi so that subgroups that are resistant to one MOA are controlled by the fungicide partner with a different MOA.
3. Rotate fungicides with different mechanisms of action, not just different label names. Rotating fungicides involves alternating products so that you avoid back-to-back treatments with any one single-site fungicide or fungicides with a medium to high risk of resistance. Neither fungicide combinations nor alternating prevent the selection of resistance. The best either approach can do is slow the process. Carefully read fungicide labels to determine if any fungicides cannot be mixed or rotated together. There are often limits on the total amount to be applied, the number of allowable applications per season, and restrictions on back-to-back applications.

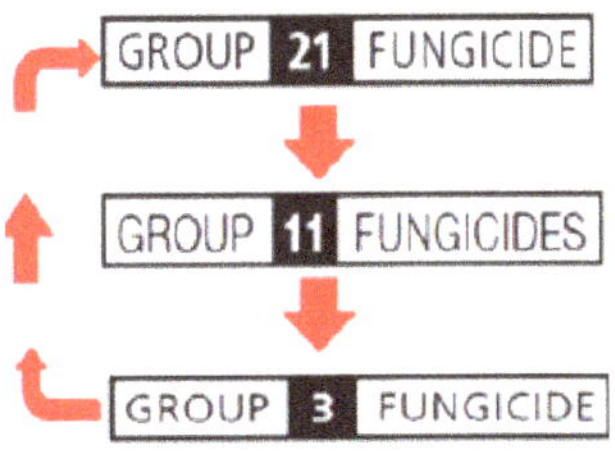

Fig. 6

Fig. 7 Wheat Stripe Rust

4. **Make use of disease predictive models and scout fields** frequently to effectively time applications of fungicides. The most critical time to use at-risk fungicides is early in an epidemic when the pathogen population is small. Similarly, a preventive spray program is less risky than a rescue program because selection pressure is applied to fewer individuals. Use multisite fungicides in later parts of the growing season when disease pressure is often greatest. Pathogens that have multiple generations in a growing season, have a high reproductive rate, and are aerially dispersed are more prone to develop resistance and become a problem. In contrast, soilborne pathogens produce fewer offspring per season than their foliar counterparts. Where an at-risk fungicide is used for soilborne disease control, resistance development is likely to be slow because comparatively few individuals are exposed to selection pressure.
   a. Apply labeled rates with adequate coverage of the foliage to produce good disease control.
   b. If applicable, use a seed or soil fungicide treatment that requires only one or two applications per season, thus keeping the risk of resistance low.
   c. Following application, monitor the field to determine how well the fungicide worked and for signs of any failure.
5. 5. Manage fields, farms, or sites wisely:

   1. elect resistant varieties of plants/crops that are less susceptible to important pathogens. Pathogens reproduce at higher rates on susceptible varieties compared with resistant or partially resistant varieties.
6. Capitalize on cultural practices to reduce the pathogen population or create conditions less favorable for the pathogen to develop. When fewer non-chemical control methods (such as sanitation) are used, resistance is more likely to develop, or will develop more rapidly.
7. Monitor and maintain agronomic factors such as soil moisture and crop nutrition, and manage other pests, to avoid crop stress throughout the season.
   - Inadequate or excessive fertilization with nitrogen may increase disease incidence in some crops. For example, early blight of potato and tomato and dollar spot of turfgrass are favored by nitrogen deficiency. Alternatively, the severity of spring dead spot of bermudagrass and some foliar diseases of wheat is increased with intensive nitrogen fertilization.

   - Excessive irrigation or frequent irrigation with small amounts of water increases the incidence of many diseases by promoting spread of the pathogen and extending periods of leaf wetness, and high soil moisture.
8. **Rotate crops** to allow greater flexibility in choosing different tillage practices and/or the use of fungicides with different MOAs. Be aware that applying a fungicide unnecessarily to a crop in one year in the absence of fungus could

result in undesirable effects on pathogens of rotational crops. NOTE: It is possible to rotate different crops between years and, yet, still use the same fungicide MOA to control the same disease. When one crop can serve as a source of inoculum for a subsequent crop, the alternation scheme among at-risk fungicides should be continued between successive crops such that the first at-risk fungicide applied to a crop belongs to a different group than the last at-risk fungicide applied to the previous crop.

9. Be aware that resistant plant pathogens can spread from nearby fields, highways, railroads, or utility rights-of-way areas near your farm or treated site. Local and regional cooperation in resistance management is essential. Participate in Extension programs that conduct area-wide monitoring for the presence of disease pathogens and potential resistance development. Submit reports of fungicide control failures so that the possibility of resistance can be monitored and evaluated.

FUNGICIDE RESISTANCE STEWARDSHIP

The introduction of the SDHI group of fungicide chemistry has been very welcome over recent seasons and has brought extra disease control and yield increases in wheat and barley crops. It was always flagged as being at moderate or high risk of fungicide resistance and sadly over the last two seasons we have seen the first examples of cereal diseases starting to adapt to the chemistry.

Fungicide resistance development is very like the much publicised development of hospital superbugs where antibiotics are overused. The driving principle is very Darwinian and all about the survival of the fittest. Where fungicides are widely used there is a competitive advantage to any individuals in a disease population which carry mutations that make them better able to survive (fitter) compared to the original 'wild' population.

These mutated individuals thrive and prosper and soon the sensitive population is overrun and the new, less sensitive part of the population becomes widespread. This can be rapid and complete – as was seen when we first picked up resistance in septoria to the strobilurins. Within a season this group was all but ineffective against that disease. Or resistance development can be slow, in the manner seen with the azole group of chemistry where we have seen a slow erosion but have been able to retain them in septoria fungicide programmes. SDHI resistance is likely to sit somewhere between these two examples and so strong stewardship to keep them for as long as we can in as effective a form as possible is really important.

It is particularly topical for this season as just before Christmas there were confirmed examples of reduced resistance in septoria sampled from UK crops. Some of these isolates were 100-fold less sensitive compared to the original population which sounds dramatic. It clearly isn't a good news story but to set it in context we are quite used to isolates of septoria that are 500 or even a 1000-fold less sensitive to azoles. And of course it comes on top of detections of mutants of reduced sensitivity in net blotch in barley.

Good stewardship is a mix of regulatory restrictions and more voluntary good practice in the field. We know that the SDHIs are at risk and that azole fungicides have been declining gradually in efficacy for years so the best strategy now is to use both groups in as balanced and as sensible a way as possible to try to delay any sudden declines in efficacy that would be hastened by high risk practices such as the use of straight products or very unbalanced mixtures.

Multi-site fungicides have a very important role to play in protecting both groups – and ideally we would be growing varieties with better resistance too. The statement below is the guidance on SDHI stewardship prepared by the Fungicide Resistance Action Group–UK which SRUC contributes to through Scottish Government RESAS funding.

SDHI Anti-resistance strategies

Scenario 1 – SDHI Foliar applications

1. Follow the statutory requirement to limit the number of applications to two SDHI fungicide-containing sprays per crop.
2. Always use SDHI fungicides in mixture with at least one fungicide from an alternative mode of action group which has comparable efficacy against the target pathogen(s).
3. Tank mixing two SDHI fungicides is not an anti-resistance strategy. In any tank mix the SDHI should be applied in a balanced mixture with at least one fungicide with comparable efficacy against the target pathogens from an alternative mode of action group.

Scenario 2 – SDHI Seed treatments with no efficacy against foliar pathogens

1. These do not count towards the statutory limit of two foliar SDHI applications; advice would therefore be to apply any subsequent foliar SDHI applications as described in Scenario 1 above.
2. The SDHI seed treatment should be co-formulated with a fungicide with an alternative mode of action to reduce selection pressure on seed-borne pathogens.

Scenario 3 – SDHI Seed treatments with efficacy against foliar pathogens

1. There are currently no SDHI seed treatments with efficacy against foliar pathogens approved in the UK
2. Should any such seed treatments be approved, FRAG-UK would advise that these should count as one of the statutory limit of two SDHI applications to a crop. It would therefore become necessary that growers keep records of seed treatments applied.

Fungicide Resistance and the FRAC Code

Fungicide resistance is a very real concern in field crops in Tennessee and as fungicide resistance continues to be reported, it becomes equally important to be aware of the FRAC code(s) of the fungicide being applied. When a fungal plant pathogen becomes resistant to a specific fungicide, then the pathogen is either less or completely insensitive to fungicides in that FRAC group. The FRAC group or code can usually be found in the upper corner of a fungicide label (see the encircled numbers in the fungicide labels below - Fig 8).

The FRAC code is a number and/or letter combination assigned by the fungicide resistance action committee (FRAC) to group together active ingredients which demonstrate potential for cross resistance. Fungicides with the same FRAC code are at risk for cross resistance because they have the same target site. The target site and mode of action of a pesticide are sometimes incorrectly used interchangeably. Mode of action refers to the biochemical pathway being targeted within the pathogen. The target site is the specific process and/or enzyme which the fungicide active ingredient is interfering with.

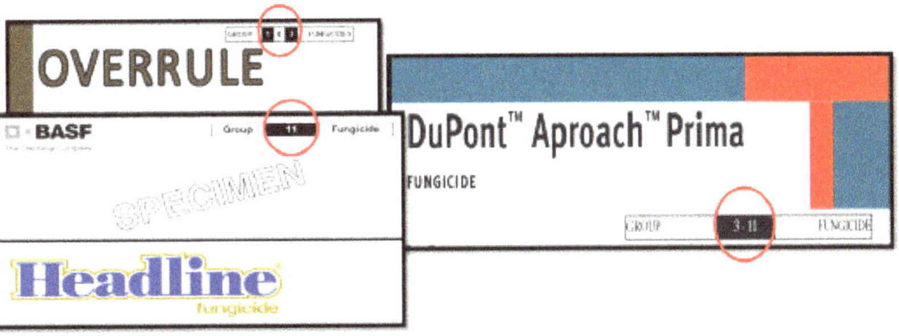

Fig. 8 The FRAC group or code can usually be found in the upper corner of a fungicide label, circled in red on images.

Fungicides may have the same mode of action but different target sites which results in a different FRAC code. For example (Table 2), QoI/Strobilurin fungicides and SDHI fungicides both inhibit fungal respiration, but they target a different part of the respiratory system.

To determine fungicide resistance risk (Low, Medium, or High) not only is the FRAC group of the chemical compound taken into account, but also how it interacts within the pathogen, and how the pathogen responds. Some pathogen-fungicide combinations demonstrate higher risk of resistance than others. There are many pathogens that have become resistant to fungicides.

A fungicide application may have no effect at preventing or inhibiting fungal growth (disease development) in a field where the pathogen population is predominantly resistant to that fungicide. Proper fungicide use will help prolong the effectiveness of these fungicides against other still-sensitive pathogens.

To receive the maximum benefits from a fungicide application, it is best to apply fungicides according to label recommendations and only when the pathogen pressure is high enough to warrant treatment. Repeated uses of a fungicide with a single FRAC code are not typically recommended. If disease severity continues to increase after fungicide application, it may be possible that the pathogen is resistant to fungicides with that FRAC code, and continued applications of that fungicide would be ineffective.

For example, there have been instances where frogeye leaf spot disease continues to progress in soybean after an application of a solo QoI/Strobilurin (FRAC code 11) product. Samples from such fields have confirmed that the population is dominated by QoI/Strobilurin resistant isolates of the pathogen. In such situations a DMI/Triazole (FRAC code 3), MBC (FRAC code 1), or a combination product would be a more effective choice to manage frogeye leaf spot.

Fungicide mixtures that combine two or more FRAC codes provide additional security that one of the active ingredients will kill the pathogen even if it is resistant to the other active ingredient in the mixture. This reduces the ability of fungicide-resistant pathogens to reproduce and populate the field, thus maintaining efficacy. However, if the pathogen is already resistant to one component of a fungicide mixture, using a fungicide mixture may eventually select for resistance to the other component. Be aware that chemical treatments are not the only disease management option. Planting resistant varieties is the first line of defense against many pathogens, and crop rotation may also help reduce disease.

FRAC codes help in fungicide resistance management

Code numbers on fungicide labels, called "FRAC" groups, can help you develop chemical rotations that delay fungicide-resistance in vegetable diseases. Fungicide resistance occurs when repeated applications of a related class of fungicides kill susceptible strains, selectively allowing the survival and reproduction of resistant ones. The ultimate result is that a formerly effective product may no longer work, which is a major problem given the ability of diseases such as *Phytophora* and powdery mildew to devastate Michigan vegetables.

GROUP	11	FUNGICIDE

Fig. 9 FRAC codes typically appear in a text box on the upper right corner of the first page of fungicide labels.

Rotation of fungicides with different modes of action is a key component of resistance management and can help maintain good products in your chemical toolbox. However, it is not enough to simply use products with different active ingredients as they could still be related and have the same mode of action. For example, Cabrio and Quadris contain different active ingredients – pyraclostrobin and azoxystrobin, respectively – yet both are in the strobilurin family and attack fungi by the same mechanism.

For this reason, the Fungicide Resistance Action Committee (FRAC) developed a numerical coding system to help you identify products that attack fungi in distinctly different ways. (A similar system has been developed for insecticides.) Simply put, products assigned the same number act the same way, while products that do not share the same numerical group act differently. For example, Cabrio and Quadris are both in FRAC group 11 and reduce the ability of fungi to convert food to energy. However, Folicur, with FRAC-code of 3, acts differently; it is a DMI-fungicide that inhibits production of a critical building block of fungal cell membranes. Rotating among fungicides with different FRAC numbers will help delay development of resistance to any one mode of action.

How can you apply FRAC codes in planning your own fungicide rotation? First, be sure to determine which fungicides provide good control of diseases in your vegetable crop. Michigan State University vegetable specialists, extension educators and crop consultants can help you accomplish this. Next, obtain and read labels for each fungicide and locate their FRAC codes, which are typically on the upper right portion of the first page.

Finally, plan to alternate effective fungicides with different FRAC code numbers to control problem diseases. This will mean you have to incur the additional expense of buying different products. However, in the long run, it will add years to the lives of these chemicals. For this reason, MSU vegetable researchers now make it standard practice to recommend rotation of chemicals for disease control on your farm and often report FRAC codes whenever discussing fungicide trials.

Chemical Classes

A grower should use fungicides with different modes of action. Tank-mix or rotate materials found in different fungicide families or groups to prevent or delay the development of resistant fungi. Groups below are not arranged in alphabetical order or in numerical order by group. In general, the first few groups are active on *Botrytis* and similar fungi (Groups 1, 2, 12, 9, 17 and 7). The next few groups are active on oomycetes (groups 4, P7, 40, 49, 22, 27, 28 and 43). Groups 3, 5, 11, U6, 50 and 13 are active on powdery mildews and many other fungi. The rest of the groups have multisite activity and make good mixing partners depending on the disease in question.

Family & Group #	Common Names	Trade Names (Combination Products)
Benzimidazole (Group 1)	benomyl	Benlate, Tersan 1991
	thiabendazole	Arbotect 20-S, Decco Salt No. 19, LSP Flowable Fungicide, Mertect 340-F
	thiophanate-methyl	Cavalier, Cleary's 3336, OHP 6672, Regal SysTec, Tee-Off,T-Methyl 4.5F AG, TM 85, Topsin M
Dicarboximide (Group 2)	iprodione	Epic 30, Ipro, Meteor, Nevado, OHP Chipco 26019, Rovral, (Interface)
	vinclozolin	Curalan, Ronilan
Phenylpyrroles (Group 12)	fludioxonil	Cannonball, Emblem, Maxim, Medallion, Mozart, Scholar, Spirato, (Academy, Miravis Prime, Palladium, Switch)
Anilinopyrimidines (Group 9)	cyprodinil	Vangard (Palladium, Switch, Inspire Super)
	pyrimethanil	Penbotec, Scala, (Luna Tranquility)

Family & Group #	Common Names	Trade Names (Combination Products)
Hydroxyanilide (Group 17)	fenhexamid	Decree, Elevate, Judge
	fenpyrazamine	Protexio
Carboxamide (Group 7)	boscalid	Emerald, Endura, (Encartis, Honor, Pageant, Pristine)
	carboxin	Vitavax
	fluopyram	Luna Privilege, Velum Prime (Broadform, Luna Experience, Luna Sensation, Luna Tranquility, Propulse)
	flutolanil	Contrast, Moncut, ProStar
	fluxapyroxad	(Lexicon, Merivon, Orkestra)
	inpyrfluxam	Excalia
	isofetamid	Kenja
	oxycarboxin	Carboject, Plantvax
	penthiopyrad	Fontelis, Velista, Vertisan
	pydiflumetofen	Miravis, Posterity, Miravis Ace A (Miravis Neo, Miravis Prime, Miravis Duo, Miravis Top)
	solatenol	Aprovia (Contend A, Elatus, Mural)
	(benzovindiflupyr)	
Phenylamide (Group 4)	mefenoxam	Apron, Ridomil Gold, Subdue MAXX, (Quadris Ridomil Gold, Uniform)
	metalaxyl	Acquire, Allegiance, MetaStar, Ridomil, Sebring, Subdue
	oxadixyl	Anchor
Phosphonate (Group P7)	aluminum tris	Aliette, Flanker, Legion, Signature, Areca
	Phosphorous Acid	Agri-Fos, Alude, Appear, Fiata, Fosphite, Phospho Jet, Phostrol, Rampart, Reload
Cinnamic acid (Group 40)	dimethomorph	Forum, Stature, (Orvego, Zampro)
	mandipropamid	Micora, Revus, (Revus Top)
OSBPI (Group 49)	oxathiapiprolin	Segovis (Orondis Gold)
Triazoles carboxamide (Group 22)	ethaboxam	V-10208
Group 27	cymoxanil	Curzate, (Tanos)
Carbamate (Group 28)	propamocarb	Banol, Previcur, Proplant, Tattoo
Benzamide (Group 43)	fluopicolide	Adorn, Presidio
Demethylation-inhibiting (Group 3)		
Piperazines	triforine	Funginex, Triforine
Pyrimidines	fenarimol	Focus, Rubigan, Vintage
Imidazole	imazalil	Fungaflor, (Raxil MD Extra)
	triflumizole	Procure, Terraguard, Trionic
Triazoles	cyproconizole	Sentinel
	difenoconazole	Dividend, Inspire, (Academy, Briskway, Contend A, Inspire Super, Quadris Top, Revus Top) Miravis Duo

Family & Group #	Common Names	Trade Names (Combination Products)
	fenbuconazole	Enable, Indar
	flutriafol	Topguard, (Topguard EQ)
	mefentrifluconazole	Avelyo, Cevya, Maxtima (Navicon)
	metconazole	Quash, Tourney
	ipconazole	Rancona
	myclobutanil	Eagle, Hoist, Immunox, Laredo, Nova, Rally, Sonoma, Systhane
	propiconazole	Alamo, Banner, Break, Bumper, Infuse, Kestrel Mex, Miravis Ace B, PropiMax, ProPensity, Strider, Tilt, Topaz, (Aframe Plus, Concert, Contend B, Headway, Quilt Xcel, Stratego)
	prothioconazole	Densicor, Proline (Propulse)
	tebuconazole	Bayer Advanced, Elite, Folicur, Lynx, Mirage, Orius, Raxil, Sativa, Tebucon, Tebuject, Tebusha, Tebustar, Toledo, (Absolute, Luna Experience, Unicorn), etc.
	tetraconazole	Mettle
	triadimefon	Bayleton, Strike, (Armada, Tartan, Trigo)
	triadimenol	Baytan
	triticonazole	Charter, Trinity, (Pillar)
Morpholine (Group 5)	piperalin	Pipron
	spiroxamine	Accrue
Group U6	cyflufenamid	Torino, fastback
Group 50	metrafenone	Vivando
	pyriofenone	Prolivo, Seido
QoI Strobilurins (Group 11)	azoxystrobin	Abound, Aframe, Dynasty, Heritage, Protété, Quadris, Quilt, (Aframe Plus,Briskway, Contend B, Dexter Max, Elatus, Headway, Mural, Quadris Top,Quilt Xcel, Renown, Topguard EQ, Uniform)
	femoxadone	(Tanos)
	fenamidone	Fenstop, Reason
	fluoxastrobin	Aftershock, Disarm, Evito, Fame
	kresoxim-methyl	Cygnus, Sovran
	mandestrobin	Intuity, Pinpoint
	picoxystrobin	Aproach
	pyraclostrobin	Cabrio, Empress, Headline, Insignia, Stamina, (Honor, Lexicon, Merivon, Navicon, Orkestra, Pageant, Pillar, Pristine)
Family & Group #	**Common Names**	**Trade Names (Combination Products)**
	trifloxystrobin	Compass, Flint, Gem, (Absolute, Armada, Broadform, Interface, Luna Sensation, Stratego, Tartan, Trigo)
Quinoline (Group 13)	quinoxyfen	Quintec
Inorganic Compounds		

Coppers (Group M1)	bordeaux	None
	copper ammonium	Copper Count-N
	complex	
	copper hydroxide	Champ, Champion, Kalmor, Kentan, Kocide, Nu-Cop
	copper oxide	Nordox
	copper oxychloride	C-O-C-S, Oxycop
	copper sulfate	Cuprofix Disperss, many others
Sulfur (Group M2)	sulfur	Cosavet, Kumulus, Microthiol Disperss,Thiosperse
Lime sulfur	Ca polysulfides	Lime Sulfur, Sulforix
Ethylenebisdithiocarbamates	mancozeb	Dithane, Fore, Penncozeb, Protect,
(EBDC) (Group M3)		Manex, Manzate, Roper, Wingman, (Dexter Max, Gavel)
	maneb	Maneb
	metiram	Polyram
EBDC-like (Group M3)	ferbam	Carbamate, Ferbam
	thiram	Difiant, Spotrete, Thiram
	ziram	Ziram
Aromatic Hydrocarbon (Group 14)	dicloran (DCNA)	Allisan, Botran
	etridizole	Terrazole, Truban
	pentachloronitro-	Autilus, Defend, Engage, PCNB,
	benzene	Terraclor, (Premion)
Chloronitrile (Group M5)	chlorothalonil	Bravo, Daconil, Docket, Echo, Ensign, Exotherm Termil, Funginil, Legend,Manicure, Pegasus, Terranil, (Concert, Spectro)
Phthalimides (Group M4)	captan	Captan
Guanidines (Group U12)	dodine	Syllit
QiI fungicides (Group 21)	cyazofamid	Ranman, Segway
Polyoxin (Group 19)	polyoxin	Affirm, Endorse, Oso, Ph-D, Tavano, Veranda
Group 29	fluazinam	Omega, Secure
Thiazolidine (U13)	flutianil	Gatten

BASED ON CHEMICAL COMPOSITION

The chemical available for plant disease control runs into hundreds, however, all are not equally safe, effective and popular.Major group of fungicides used include salts of toxic metals and organic acids, organic compounds of sulphur and mercury, quinines and heterocyclic nitrogenous compounds. Copper, mercury, zinc, tin and nickel are some of the metals used as base for inorganic and organic fungicides. The non metal substances include, sulphur, chlorine, phosphorous etc. The fungicides can be broudly grouped as follows and discussed in detail.

Groups of Fungicides – Copper Fungicides, Sulphur Fungicides and Mercury Fungicides Copper Fungicides

The fungicidal action of copper was mentioned as early as 1807 by Prevost against wheat bunt disease (Tilletia caries), but its large scale use as a fungicide started in 1885 after the discovery of Bordeaux mixture by Millardet in France. The mixture of copper sulphate and lime was effective in controlling downy mildew of grapevine caused by Plasmopara viticola and later, late blight of potato (Phytophthora infestans).

Some other copper sulphate preparations later developed were Borduaux paste, Burgandy mixture and Cheshnut compound which are all very effectively used in the control of several plant diseases. In addition some preparations of copper oxy chloride preparations arev also mused. These are all insoluble copper compounds very successfully used in managing several leaf diseases and seeding diseases in nursery. Some of the important diseases controlled by copper fungicides are listed below.

COPPER SULPHATE PREPARATIONS

Boreaux Mixture

In 1882, Millardet in France (Bordeaux University) accidently observed the efficacy of the copper sulphate against the downy mildew of grapes caused by Plasmopara viticola. When copper sulphate was mixed with lime suspension, it effectively checked the disease incidence. The mixture of copper sulphate and lime was named as "Bouillie Bordelaise" (Bordeaux Mixture). The original formula developed by Millardet contains 5 lbs of $CuSO_4$ + 5lbs of lime + 50 gallons of water. The chemistry of Bordeaux mixture is complex and the suggested reaction is:

$CuSO_4 + Ca(OH)_2\ Cu(OH)_2 + CaSO_4$

The ultimate mixture contains a gelatinous precipitate of copper hydroxide and calcium sulphate, which is usually sky blue in colour. Cupric hrdroxide is the active principle and is toxic to fungal spores.

One kg of copper sulphate is powdered and dissolved in 50 litres of water. Similarly, 1 kg of lime is powdered and dissolved in another 50 litres of water. Then copper sulphate solution is slowly added to lime solution with constant stirring or alternatively, both the solutions may be poured simultaneously to a third contained and mixed well.

The ratio of copper sulphate to lime solution determines the pH of the mixture. The mixture prepared in the above said ratio gives neutral or alkaline mixture. If the quality of the used is inferior, the mixture may become acidic. If the mixture is acidic, it contains free copper which is highly phytotoxic resulting in scorching of the plants. Therefore, it is highly essential to test the presence of free copper in the mixture before applied. There are several methods to test the neutrality of the mixture, which are indicated below:

i. Field Test: Dip a well polished knife or a sickle in the mixture for few minutes. If reddish deposit appears on the knife/sickle, it indicates the acidic nature of the mixture.
ii. Litmus paper test: The colour of blue litmus paper must not change when dipped in the mixture.
iii. pH paper test : If the paper is dipped in the mixture, it should show neutral pH.
iv. Chemical test: Acid a few drops of the mixture into a test tube containing 5 ml of 10% potassium ferrocyanide. If red precipitate appears, it indicates the acidic nature of the mixture.

If the prepared mixture is in the acidic range, it can be brought to neutral or near alkaline condition by adding some more lime solution into the mixture. Bordeaux mixture preparation is cumbersome and the following precautions are needed during preparation and application.

i. The solution should be prepared in earthen or wooden or plastic vessels. Avoid using metal containers for the preparation, as it is corrosive to metallic vessels.
ii. Always copper sulphate solution should be added to the lime solution, reverse the addition leads to precipitation of copper and resulted suspension is least toxic.
iii. Bordeaux mixture should be prepared fresh every time before spraying. In case, the mixture has to be stored for a short time or a day, jaggery can be added at the rate of 100kg/100 litres of the mixture.
iv. Bordeaux mixture is sometimes phytotoxic to apples, peaches, rice varieties like IR8 and maize varieties like Ganga Hybrid 3.

Bordeaux paste

Bordeaux Paste consists of same constituents as that of Bordeaux mixture, but it is in the form of a paste as the quantity of water used is too little. It is nothing but 10 percent Bordeaux mixture and is prepared by mixing 1 kg of copper sulphate and 1 kg of lime in 10 litres of water.

The method of mixing solution is similar to that of Bordeaux mixture. It is a wound dresser and used to protect the wounded portions, cut ends of trees etc., against the infection by fungal pathogens.

Burgundy mixture

It is prepared in the same way as Bordeaux mixture, except the lime is substituted by sodium carbonate. So it is called as „Soda Bordeaux . It was developed Burgundy (France) in 1887 by Mason. The usual formula contains 1 kg of copper sulphate and 1 kg of sodium carbonate in 100 litres of water. It is a good substitute for Bordeaux mixture and used in copper-sensitive crops.

Cheshunt compound

It is compound usually prepared by mixing 2 parts of copper sulphate and 11 parts of ammonium carbonate. This formula was suggested by Bewley in the year 1921. The two salts are well powdered, mixed thoroughly and stored in a air tight container for 24 hours before being used. The ripened mixture is used by dissolving it in water at the rate of 3 g/litre. The mixture is dissolved initially in a little hot water and volume is made up with cold water and used for spraying.

COPPER CARBONATE PREPARATION

Chaubattia Paste

Chaubattia paste is another wound dressing fungicide developed by Singh in 1942 at Government Fruit Research Station, Chaubattia in the Almora district of Uttar Pradesh. It is usually prepared in glass containers or chinaware pot, by mixing 800g of copper carbonate and 800g of red lead in litre of raw linseed oil or lanolin. This paste is usually applied to pruned parts of apple, pear and peaches to control several diseases. The paste has the added advantage that it is not easily washed away by rain water.

Sulphur fungicides

Use of sulphur in plant disease control is probably the oldest one and can be classified as inorganic sulphur and organic sulphur. Inorganic sulphur is used in the form of elemental sulphur or as lime sulphur. Elemental sulphur can be either used as dust or wettable sulphur, later being more widely used in plant disease control. Sulphur is best known for its effectiveness against powdery mildew of many plants, but also effective against certain rusts, leaf blights and fruit diseases.

Sulphur fungicides emit sufficient vapour to prevent the growth of the fungal spores at a distance from the area of deposition. This is an added advantage in sulphur fungicides as compared to other fungitoxicants.

Organic compounds of sulphur are now widely used in these days. All these compounds, called as "carbamate fungicides", are derivatives of Dithiocarbamic acid, Dithiocarbamates are broadly grouped into two, based on the mechanism of action.

Dithiocarbamates

Monoalkyl Dithiocarbamates	Dialkyl Dithiocarbamates
Eg. Zineb, Maneb,	Eg. Thiram, Ziram,
Mancozeb, Nabam, Vapam	Ferbam

Mercury Fungicides

Mercury fungicides can be grouped as inorganic and organic mercury compounds. Both the groups are highly fungitoxic and were extensively used as seed treatment chemicals against seed borne diseases. Ignorance compounds show bactericidal property also.

However, due to their residual toxicity in soil and plants and their extreme toxicity nature to animal and human beings, the use of mercury fungicides is beings discouraged. In most of the countries, the use of mercury fungicides is banned and in countries like India, the use of mercury fungicides is restricted only in seed treatment for certain crops.

Heterocyclic Nitrogen Compounds, Quinones and Miscellaneous Fungicides

Heterocyclic Nitrogen Compounds

Heterocyclic nitrogen compounds are mostly used as foliage and fruits protectants. Some compounds are very effectively used as seed dressers.

Benzene compounds

Many aromatic compounds have important anti-microbial properties and have been developed as fungicides.

Quinone Fungicides

Quinone areresent naturally in plants and animals and they exhibit anti-microbial activity and some compounds are successfully developed and used in the plant disease control. Quinones are very effectively used for seed treatment and two commonly used fungicide.

Organo Tin compounds

Several other organic compounds containing tin, lead, etc. have been developed and successfully used in plant disease control. Among them, organo tin compounds are more popular and effective against many fungal diseases. These compounds also show anti bactericidal properties.

SYSTEMIC FUNGICIDES AND ANTIBIOTICS

Systemic Fungicides

Since the late 1960s there has been substantial development in systemic fungicides. Any compound capable of being freely translocated after penetrating the plant is called systemic. A systemic fungicide is defined as fungitoxic compound that controls a fungal pathogen remote from the point of application, and that can be detected and identified. Thus, a systemic fungicide could eradicate established infection and protect the new parts of the plant. Several systemic fungicides have been used as seed dressing to eliminate seed infection. These chemicals, however, have not been very successful in the cases of trees and shrubs. On the basis of chemical structure, systemic fungicides can be classified as Benzimidazoles, Thiophanates, Oxathilins and related compounds, pyrimidines, morpholines, organo-phosphorus compounds and miscellaneous group.

I. Oxathilin and related compounds

Oxathalins were the earliest developed compounds. This group of systemic fungicide is also called as carboxamides, carboxyluc acid anillides, carboxaanillides or simply as anillides which are effective only against the fungi belong to Basidiomycotina and Rhizoctonia solani. Some of the chemicals developed are (i) Carboxin (DMOC: 5,6 - dithydra-2-methyl-1, 4- oxathin-3-carboxanillide) and (ii) Oxycarboxin (DCMOD-2,3-dihydro-5-carboxanillido-6- methyl-1, 4 oxathilin-4, 4, dioxide).

III. Thiophanates

These compounds represent a new group of systemic fungicides based on thiourea. They are the derivatives of thioallophanic acid. These fungicides contain aromatic nucleus which is converted into benzimidazole ring for their activity. Hence, thiophanates are often classified under benzimidazole group and the biological activity of thiophanates resembles of benomyl.

7

USE OF COPPER-BASED FUNGICIDES IN ORGANIC FIELD CROPS

We are one of the leading and reliable manufacturer and exporter of chemical fungicides, agricultural fungicides, fungicide formulations for use in Agriculture and public health. Our company offers a wide range of fungicides and is committed to serve the world farming community with its quality products. Each and every fungicide is subjected to stringent quality control norms to ensure our large number of clientèle across the globe, get only quality products.

Fungicides are the chemical compounds used to preclude the spread of fungi or plants in crops and gardens which can campaign serious damage resulting in loss of yield. Sometimes, to fight with fungal infections fungicides are used. Two categories of fungicides are usually available, they are :

Contact fungicide:

It kills fungi when sprayed on its surface.

Systemic fungicide:

It is absorbed by the plant first and then show its effect.

Mostly fungicides are available and sold in liquid form. Some of the active ingredients present in them are sulfur, neem oil, rosemary oil, jojoba oil and the bacterium bacillus subtilis. One of the fungicide which is very dangerous to human health is Vinclozolin. Fungicides are usually used to control bacterial and fungal diseases of fruit, vegetable, nut and field crops. Some of the diseases that are controlled by using fungicides are mildew, leaf spots, blights and apple scab.

Our fungicides are highly effective specially against powdery mildew in cherries, grapes, hops, lettuce, melons, peppers and strawberries. They are just excellent crop safety profile and does not disturb any biological control agent such as predatory mites.

We are manufacturing two types of fungicides :

One is Chemical based and the other one is Organic based.

Chemical based fungicides are :

- Mancozeb
- Tricyclazole
- Carbendazim
- Hexaconazole
- Metalaxyl
- Benomyl
- Difenoconazole
- Propiconazole
- Kitazin
- Tebuconazole
- Copper oxychloride
- Copper hydroxide
- Tridemorph
- Propineb

Mancozeb

It is a grayish-yellow powder with a musty odor usually insoluble in water as well as other organic solvents. Mancozeb is a polymer of maneb combined with zinc , has low acute toxicity to mammals and is not poisonous to plants.

Tricyclazole

It is a systemic fungicide used for controlling blast disease of rice and preventing rice spike and leaf paste. Tricyclazole is a new fungicide having high effect, low toxicity , low residue and is effective to control rust, powdery mildew and anthracnose on beans .

Carbendazim

It is a systemic benzimidazole fungicide that plays a very important role in plant disease control. Carbendazim is used for controlling a wide range of pathogens and as a preservative in fruit industry, textile, paper making, leather and paint industry.

Hexaconazole

It is a systemic , broad-spectrum fungicide with protective and curative action. Hexaconazole is mainly used on bananas, cucurbits, peppers and other crops.

Metalaxyl

It is a systemic, benzenoid fungicide used in mixtures as a foliar spray for tropical and subtropical crops. Metalaxyl is also used as a soil treatment for control of soil-borne pathogens and as a seed treatment to control downy mildews.

Benomyl

It is a systemic, benzimidazole fungicide that is selectively toxic to microorganisms and invertebrates, particularly earthworms. Benomyl is used against a wide range of fungal diseases of field crops, fruits, nuts, ornamentals, mushrooms, and turf.

Difenoconazole

It is a broad-spectrum inhalation fungicide mainly used to protect and cure basidiomycetes and comycetes fungi in leaves or seeds of grape, peanut, potato, wheat and vegetables.

Propiconazole

It is a systemic foliar fungicide with protective and curative action. Propiconazole is used as a biocide against blue stain in wood preservatives and compounding with triazoles and pyrethroids in exterior varnishes.

Kitazin

Kitazin is a broad spectrum systemic fungicide belonging to the organophosphorus group for the control of various diseases such as blast and sheath blight of rice, fruit rot of chilli, early blight of tomato and potato, purple blotch of onion, anthracnose of grapes and pomegranate etc.

Tebuconazole

It is a fungicide used for the control of smuts, bunt, seed rots and seedling blights on barley, oats and wheat as a seed treatment. Also used for the control of fusarium head blight on wheat as a post-emergent treatment.

Copper oxychloride

It is a kind of fungicide which is effective and used for the control of fungus diseases of fruit trees, vines and vegetables, also helps in preventing black spot on apples, pears, tomatoes and citrus fruits .

Copper hydroxide

It is an excellent protective fungicide of blue color killing both fungi and bacteria. Copper hydroxide is widely used and applied in the agriculture industry to prevent agricultural crops from contracting diseases including copper deficiency.

Tridemorph

It is a systemic fungicide used widely over numerous crops in many countries around the world. Tridemorph is used to control the fungus Erysiphe graminis in cereals, Mycosphaerella species in bananas, and Caticum solmonicolor in tea.

Propineb

It is a basic foliar fungicide with protective action used on citrus fruit , berry fruit, rice ,tea and to control downy mildew, black rots, red fire disease, and grey mould on vines; scab and brown rot on applesand pears and leaf spot diseases on stone fruit.

The other category is **organic type of fungicides**. They are :

- Safin
- Sporrin
- Blastin
- Bio-Vitrioll

Safin

We produce two varities of Safin, one is an organic derived from natural extracts of Allicin & Calotropin . The other one is 6G Safin derived from 6%granules. Safin is toxic free and control wide range of fungi and nematodes with a long standing period of action. For more information, click on the lonk below:

Sporrin

There are two types of Sporrin manufactured by us , one is a natural organic derived from natural extracts of triterpenes & alkaloids . The other one is Sporrin 6G derived from 6% granules. Sporrin is a potential fungicide & bactericide used for controlling diseases like powdery mildew, downy mildew, rot, anthracnose, wilt, leaf spot, cankers, early & late blight etc.

Blastin

Two categories of Blastin are produced by us. One is a natural extract of monoterpene, G-terpinene & Alkaloids and the other one is Blastin 6G derived from 6% granules.

Table 1. Classes of Fungicides, with Examples

Class of Fungicide	Examples
Substituted Benzenes	Chloroneb, chlorothalanil, Hexachlorobenzene, pentachloronitrobenzene
Thiocarbamates	Ferbam, metam sodium, thiram, ziram
Ethylene Bis Dithiocarbamates (EBDC's)	Mancozeb, maneb, nabam, zineb
Thiophthalimides	Captan, captafol, folpet

Class of Fungicide	Examples
Copper compounds	
Organomercury compounds	Ethyl mercury, methyl mercury, phenyl mercuric acetate
Organotin compounds	Fentin, triphenyl tin
Cadmium compounds	
Miscellaneous organic fungicides	Benomyl, cyclohexamide, iprodione, metalaxyl, thiabendazole, triadimefon

The benefits of fungicide use have been many. In agriculture, fungicides control pests that may rob water and nutrients from crop plants or may cause food spoilage as the products are brought to market. Fungicides may also prevent the growth of fungi that produce toxins, such as aflatoxins. Fungicides also have important industrial applications and are important in preserving the purity and safety of certain pharmaceutical agents.

In 1997 there were an estimated $0.8 billion in sales of fungicides in the United States, about 7 percent of the total pesticide market. In 1997, worldwide, 5.7 billion pounds of pesticides were used, of which 0.5 billion were fungicides. Of the1.2 billion pounds of conventional pesticides used in the United States in 1997, a total of 81 million pounds of fungicides were used; 79 percent of the use was in agriculture. Generally, the United States has experienced a downward trend in total fungicide use since 1970.

There are numerous classes of fungicides, with different modes of action as well as different potentials for adverse effect on health and the environment (see Table 1). Most fungicides can cause acute toxicity, and some cause chronic toxicity as well. Hexachlorobenzene, now banned or severely restricted in most parts of the world, has been associated with human poisoning from contaminated seed grain and poisoning of infants from misuse in laundry solutions. Metam sodium and other thiocarbanates are skin irritants that can cause reactive airway disease at low doses and severe toxicity and even death at high doses. The ethylene bis dithiocarbamates (EBCDs) are suspected human carcinogens and are tightly regulated in the United States.

Organic mercurials have caused severe acute and chronic toxicity. Worldwide, there have been a number of incidents of treated seed grain fed to people, with disastrous consequences in terms of acute poisoning and damage to fetuses. Phenyl mercuric acetate is no longer used as a paint preservative in the United States because it off-gases elemental mercury into the air, with the potential for causing toxicity to young children. Organotin compounds also have serious human toxicity and are very toxic to the environment; their use is banned or severely restricted in most of the world. Likewise, due to human toxicity concerns, cadmium is no longer used as a fungicide in the United States.

COPPER-BASED FUNGICIDES IN ORGANIC AGRICULTURE

The use of copper fungicides to combat plant diseases has a long tradition, dating back to the late 19th century, when Alexis Millardet accidentally discovered the efficacy of neutralized copper sulfate to reduce grapevine downy mildew. This discovery revolutionized agricultural production by providing the first effective tool to control various phytopathogens. Although many fungicidal active substances have been discovered in the meantime, copper-based plant-protection products are still widely used in organic and conventional agriculture, due to their many advantages and the lack of alternatives.

One of the main advantages of copper is its wide spectrum of activity against bacteria, oomycetes, ascomycetes, and basidiomycetes, including diseases of worldwide importance, such as downy mildew of grape and late blight of potato, but also secondary or minor diseases in minor crops, such as vegetables or specialty fruit. Other favorable agronomical properties of copper include high efficacy under conditions of rain, its multisite mechanism of action that minimizes the risk of development of resistant pathogen strains, the relatively low acute toxicity for terrestrial vertebrates, and the low costs.

Besides its use as a fungicide, copper is also authorized as a micronutrient leaf fertilizer and widely used in feed additives. Micronutrient fertilizers may contain copper under the form of copper salt, copper oxide, copper hydroxide, copper chelate, and copper oxychloride, either as a copper complex or as mixtures of components (i.e., copper-based fertilizer and copper fertilizer solution). An optimal supply range for copper in plant nutrition is 0.01–0.02 mg g^{-1} dry weight measured as plant content, but requirements may depend on the crop or even the cultivar.

Most of the copper used as a feed additive ends up in the manure, which is considered a substantial source of copper input in arable crops and fodder production, especially when pig slurry and manures from conventional farms are imported in organic farms, but it has no relevance in the control of soil-borne pathogens. Being an element having scarce mobility in soil, repeated foliar applications of copper-based plant-protection products lead to copper accumulation in the soils and to potentially consequent negative impacts on soil fertility, even though studies put the negative effects in perspective.

The maximum copper quantity allowed in plant protection has been successively restricted in Europe over the last decades and is currently limited by the European plant-protection legislation to a maximum of 28 kg per ha over a period of 7 years (status 2021) (regulation (EU) 2018/1981). The final objective would be to phase copper out, as it is included in the list of candidates for substitution in the EU (Part E of the Annex to Regulation 540/2011) and Switzerland (Aktionsplan Pflanzenschutzmittel, www.BLW.admin.ch, accessed on 7 March 2022). In organic farming, copper-based plant-protection products and fertilizers are explicitly allowed according to the European Commission implementing Regulation (EC) No. 2021/1165 .

In addition to European legislation, EU Member States (MS) can further restrict the quantity or authorize only specific uses. For example, in five out of 27 EU member states (Denmark, Sweden, Finland, the Netherlands, and Estonia), copper is not registered as a plant-protection product. Furthermore, in a range of countries, farmer organizations self-restrict copper use beyond the legal requirements. As a consequence of overlapping EU regulations and of differences in nationally authorized plant-protection products and practices at different levels, it is difficult to get a precise overview of permitted uses (i.e., in which crops against which pathogens in which maximum quantity).

There is a substantial difference between the maximum quantity of copper permitted for a specific use (crop/disease) and its real use, as already highlighted in Switzerland, France, and Germany. For example, a survey in Switzerland found that organic farmers use, on average, between 3 and 80% of the maximal permitted amount of copper depending on the crop, indicating a high awareness among farmers of the need to reduce copper use. Katsoulas et al. report considerable variability in copper use between regions and/or producers in European organic farming, depending, for example, on climatic conditions, cultivars, or production systems, and they indicate that annual copper limits were not always respected.

Even though various tools and strategies to reduce copper-based plant-protection products, such as resistant varieties, cultural measures, decision-support systems, and alternative plant-protection products, are available for farmers, completely abandoning copper fungicides would lead to high yield losses in many crops at the current time. To develop strategies and pathways for a complete phasing out of copper-based plant-protection products, policymakers need information on the current dependency on copper-based plant-protection products, the implementation of alternative strategies, and the need for copper alternatives in European organic-crop production. Our survey aimed at

i. giving an overview on the current legal status of copper uses in organic farming in different European countries,
ii. the discovery of the main copper-consuming crops, and
iii. an estimate of the total amounts of copper consumed in European organic crop protection.

The survey is based on knowledge of national experts (researchers, extension specialists, and experts from farmers' associations) from 12 European countries, including Belgium, Bulgaria, Denmark, Estonia, France, Germany, Hungary, Italy, Norway, Spain, Switzerland, and the UK.

Countries Included in the Study

The present survey on the theoretically allowed and real use of copper-based plant-protection products in European organic agriculture includes 12 European countries, covering all European agroclimatic zones as defined by the European Plant Protection Organisation (EPPO).

Zones included are the Mediterranean (Italy, Spain, and part of France), Maritime (Belgium, part of France, Germany, Switzerland, United Kingdom, Denmark, and part of Norway), North East (Estonia and part of Norway), and Central (Bulgaria and Hungary) zones.

Organically Managed Area

The overall organically managed area and the organically managed area per crop for each country were obtained from Statistics.FiBL.org (accessed on 7 March 2022), which gives access to the data collected annually by the Research Institute of Organic Agriculture (FiBL) and partners. Only selected crops were listed, and unspecific categories were excluded. The sum of the area for the selected crops does, therefore, not correspond with the total organic farmland for a country.

Permitted and Real Copper Use

The European pesticide database was consulted for approval of copper compounds as active substances in the involved countries and amended by information on emergency approvals at the national level. National databases (e.g., listed by the European and Mediterranean Plant Protection Organisation EPPO) were checked for permission of copper-based plant-protection products and permitted uses (crops, dosages in kg ha^{-1}).

This information was further verified and supplemented by national experts from research institutes, farmers' associations or extension services participating in the EU project on which this survey is based, or by experts in their network, amounting to 29 involved national experts.

This expertise was included to fill knowledge gaps, as plant-protection products are also allowed for use by way of emergency registrations, minor use registrations, or special national instruments, such as Off-Label Extension of Authorisation for Minor uses, and this is not mapped in the EU pesticide database.

For key crops, a description of the most widely used crop-protection strategies, targeted on diseases controlled by copper, was compiled based on the knowhow of national experts of relevant countries. Moreover, the immediate potential for further reduction of copper use—which does not compromise yield stability—was assessed.

Calculations and Statistics

The amount of copper allowed in organic plant protection was calculated for each crop and country as follows:

$$\text{Copper}_{\text{allowed}}\ (\text{kg y}^{-1}) = \text{Area (ha)} \times \text{permitted use (kg ha}^{-1}\ \text{y}^{-1})$$

The amount of copper really used in organic plant protection was calculated for each crop and country as follows:

$$\text{Copper}_{\text{used}}\ (\text{kg y}^{-1}) = \text{Area (ha)} \times \text{estimated real use (kg ha}^{-1}\ \text{y}^{-1})$$

The amount of copper really used by organic farmers was put into perspective with the allowed amount for each crop and country as follows:

Copper utilization rate = $(\text{Copper}_{\text{used}}) \times (\text{Copper}_{\text{allowed}})^{-1}$

The Pearson correlation coefficient was calculated by using the software IBM SPSS Statistics for Windows (IBM Corp. Released 2015. Version 23.0. Armonk, NY, USA: IBM Corp.).

Results

Authorized Uses of Copper in European Agriculture

While copper compounds are approved as active substances for plant protection at EU level, plant-protection products containing copper are registered for the use in individual crops and against individual pathogens at the national level, resulting in different copper limits for individual cases (countries × crop). Our survey shows that copper-based plant-protection products are registered in 25 out of 30 European countries (22 EU member states, as well as the non-EU countries, namely Norway, Switzerland, and the United Kingdom). In five out of 27 EU member states, copper is not registered as an active substance (Denmark, Sweden, Finland, the Netherlands, and Estonia). The copper compounds authorized as active substances in fungicides in the EU are Bordeaux mixture, copper hydroxide, copper oxide, copper oxychloride, and tribasic copper sulfate. However, not every copper compound is registered in each country. Copper hydroxide and oxychloride are both registered in 22 countries, followed by tribasic copper sulfate (16 countries), copper oxide (12 countries), and Bordeaux mixture (ten countries).

Different overlapping European and national regulations, the lack of centralized databases, and additional self-restrictions beyond the legal requirements by national label organizations make it difficult to get an overview on allowed copper uses (country/crops/pathogens) and the limits that apply. At the international level, Demeter International limits the amount averaged over 7 years to 3 kg ha^{-1} y^{-1} and recommends a maximum of 500 g/ha/spray. The Demeter International standard allows for exemptions in wine- and hop-growing regions with high fungal pressure. In these cases, the respective certifying organization may grant an exemption for the use of an average amount of up to 4 kg ha^{-1} y^{-1} over 5 years for grapes and hops. At the national level, associations such as Bio Austria (AT), Bioland (DE), Naturland (DE), Bio Suisse (CH), or PRO-BIO (CZ) defined crop-specific limitations.

In fertilizers containing the macronutrients ammonium (N), phosphorous (P), potassium (K), magnesium (Mg), and calcium (Ca), copper is considered a contaminant, and maximum levels are defined. Here, the copper content in an organic fertilizer must not exceed 300 mg kg^{-1} dry matter by Regulation (EU) 1009/2019. However, we found that fertilizers do not undergo stringent approval procedures in many European countries, and the full composition is rarely divulged by the manufacturer. Furthermore, leaf fertilizers may contain substances with well-known plant-protection properties, such as copper; phosphonate; or even insecticides, such as Matrine. Harmonization is intended on the European level for the future, as the new Fertilizer Regulation offers optional harmonization: In order to get the CE marking, the requirements of the new regulation have to be met. For their internal markets, the EU member states will still have the possibility to set specific rules, and it is still possible to recognize products mutually.

Beyond European regulation, an increasing number of organic farmers' associations and certifiers have joined forces to develop a harmonized and transparent inputs assessment scheme to facilitate use of high-quality products in organic farming. At present, the network publishes input lists for Austria (AT), Croatia (HR), Germany (DE), Italy (IT), Netherlands (NL), Sweden (SE), and Switzerland (CH).

Copper Use by European Organic Farmers

Copper Use Authorization and Estimated Real Use Per Hectare and Year

This survey in the 12 selected European countries covers 2.9 million hectares of organically managed horticultural (temperate fruits; grapes; olives; nuts; and vegetables, including potato) and arable crops. Permanent grasslands were not included. The use of copper is permitted on a total 1.55 million hectares. Among the countries included in the survey, Italy, Spain, and France represent the largest areas of horticultural crops in organic production in Europe (75%). Together with the other nine countries included in this survey, 83% of the organically managed European (EU + Switzerland + UK + Norway) horticultural area is covered.

Copper as a plant-protection product is permitted on a wide range of perennial, as well as on selected vegetable, crops in the 12 surveyed European countries. In arable crops, copper is only used on potato as a plant-protection product, but it is sometimes used as a leaf fertilizer, without expectations regarding disease control. The number of uses allowed per country depends directly on whether the climate is suitable for certain crops.

In Italy, Spain, and France, copper is allowed in more than 40 crops/crop categories. In the Southern Central European countries, such as Germany or Switzerland, the use is permitted on approximately 20 crops, whereas in the Northern Central European countries, five to ten uses are registered. In the examined countries, copper is mainly allowed on apple and grapevine (12 countries), pear (eleven), potatoes and plums (eight each), and cherries and strawberries (seven each).

Some crops, such as olives or almonds, are only grown in a few of the surveyed countries; thus, consequently, only a few allowed uses are registered. Permitted amounts of copper vary between 1.2 and 4 kg ha^{-1}, but they are between 3 and 4 kg ha^{-1} y^{-1} in most cases (crops × country). In Italy and France, in all permitted cases, 4 kg ha^{-1} is allowed. In other countries (e.g., Switzerland, Germany, and Spain), permitted amounts vary depending on the crop.

Organic farmers use between 0 and 100% of the maximum allowed amount of copper depending on the crop and/or country, i.e., copper utilization rates ($copper_{used}$) × ($copper_{allowed}$)$^{-1}$) ranged between 0 and 1. Over all surveyed countries, copper-utilization rates were <0.25 in 27% of the allowed cases (crops × country), <0.5 in 56%, and >0.75 in 27% of the allowed cases, with some variability between countries. The UK showed the highest percentage of cases with utilization rates < 0.5 (94%), followed by Germany (70% of cases), Switzerland (67%), Spain (62%), France (61%), Bulgaria (47%), and Italy (41%). The relatively high percentage of cases with high copper-utilization rates in Belgium, Norway, and Hungary has to be put into perspective with low maximum permitted copper quantities (2–2.6 kg ha^{-1} y^{-1} in Belgium for all crops except potatoes; 3 kg ha^{-1} y^{-1} in Norway and Hungary) in most crops and, for Norway, with a very limited number of permitted uses (3 crops).

Copper utilization rates vary between crops. For example, in apples, 3–4 kg ha^{-1} y^{-1} is permitted in Germany, Spain, Switzerland, or the UK, but, on average, only 2–1 kg is used. In other crops, limits are often fully exploited, even though differences between countries may exist. For example, in grapevine, the maximum allowed amount of copper is used in five out of the eight countries which allow its use.

Annual Copper Use

In order to put copper use in organic farming into perspective with its impact on the economy and the environment, copper consumption was calculated for each case (crop × country) by multiplying the amount of copper allowed/used (kg ha^{-1}) by the corresponding cultivated area (ha). Overall, organic farmers in the 12 surveyed countries made use of about 53% of the total authorized amount of copper, equaling to approximately 3 300 t of copper metal per year.

Among the countries included in this survey, the highest amounts of copper per year were used in Italy (approximately 1 550 t), Spain (approximately 1100 t), and France (approximately 500 t), followed by Bulgaria (67 t), Germany (42 t), and Hungary (22 t). Copper use was highly correlated to the organically managed area on which copper use is allowed ($R^2 = 0.941$, $p < 0.01$, Pearson), which obviously depends on the size of the country, the share of organically managed land, the cultivated crops, and the copper authorizations in these crops.

In the 12 surveyed countries, olive production with an estimated 1263 t, grapevine production with 990 t, and almonds with 317 t account together for nearly 80% of the estimated annual copper use. Nuts, apples, vegetables (fruit), oranges, hazelnuts, vegetables (no details), potatoes, and barley (as fertilizer) each use between 40 and 80 t per year, accounting together for 14% of total copper consumption. Protein crops, lemons and lime, cherries, citrus (no detail), kiwis, brassicas, wheat (as fertilizer), apricots, plums, pears, and peaches each use between 10 and 25 t per year, accounting together for 6% of total annual copper use. All other crops/categories use less than 10 t per year. After aggregating crops into crop categories, "vegetable crops" (aggregating "vegetables fruit" and "vegetables no details") rank fourth (106 t y^{-1}, 3.3%) after the top three categories, namely "olives" (1 263 t y^{-1}, 38.8%), "grapes" (990 t y^{-1}, 30.4%), and "nuts" (aggregating "almonds", "hazelnuts", "nuts no details", and "walnuts") (459 t y^{-1}, 14%). "Citrus fruit" ranks fifth (102 t y^{-1}, 3.1%), before "pome fruit" (78 t y^{-1}, 2.4%), "cereals" (72 t y^{-1}, 2.2%, fertilizer use), and "stone fruit" (67 t y^{-1}, 2.1%) in the 12 surveyed countries. Potatoes only ranked ninth, with 39 t y^{-1} (%1.2).

The ranking of relative importance differs substantially between countries. In Belgium, where application is allowed in seven crops, potatoes accounts for 78% of copper consumption, followed by apples (13%) and pears (5%). In Bulgaria, where 14 uses are permitted, nuts and grapevine use 69% of the total quantity, followed by cherries (9%). In France, grapevine is where the most important use is made (68%), followed by nuts (11%), vegetables (9%), apples (7%), and potatoes (2.5%), with much smaller quantities.

In Germany, copper use is relatively equally distributed between grapevine (40%), potatoes (34%), and apples (19%). In Hungary, copper is preferably used on apples (37%), nuts (28%), and grapes (23%). In Italy, an extraordinary wide range of uses is permitted. Olives (45%) and grapevine (27%) alone use up 72% of the total copper, followed by nuts and almonds (8%), citrus fruits (aggregating "lemons and limes", "oranges", and "citrus no details") (5.6%), and vegetables (2.6%).

Quite remarkably, the amount of copper used on barley and wheat (together 2.5%) as a fertilizer appears much higher than the amount used as a plant-protection product on many important crops, such as apples, cherries, apricots, peaches, pears, and plums (all < 1%). The amount of copper used in Norway with 0.5 t y^{-1} is minimal in comparison to other countries and concentrates on apples (%97) and pears (%3).

In Spain, the largest amounts of copper are used in olives (51%), almonds (23%), and grapes (20%), together accounting for 94% of the copper used. As in Italy, the use as a fertilizer in barley with 3% is higher than in plant protection on other crops, such as vegetables, oranges, apples, and nectarines (together below 2%). In Switzerland, copper use concentrates on grapevine (41%) and potatoes (32%), followed by vegetables (9%), stone fruits (6%), berries (5%), apples (4.7%), and pears (1.3%) at much lower quantities. In the UK, potatoes are the most important use (69%), followed by apples (28%) and pears (3%).

Discussion

Even though organic agriculture strictly regulates the use of inputs, organic farmers in Europe still use some inputs generally thought to be contentious (anthelmintics, antibiotics, and vitamins in livestock management; and external nutrient inputs, mineral oil pesticides, and copper-based fungicides in plant production). However, the reduction or substitution of these inputs has high priority in European agricultural policy and for organic-farming associations.

This study on the current use of copper-based fungicides was performed within the EU-funded project RELACS, in which surveys and studies related to all of the contentious inputs listed above were conducted. While this publication gives an in-depth analysis and discussion on the use of copper-based fungicides, the main findings of this survey and the surveys on the other contentious inputs investigated within RELACS have been summarized by Varga et al., putting them into a larger perspective.

For copper-based fungicides, there is a lack of data on their use, in particular, with regard to total consumption, leading to ambiguities regarding the quantities to be replaced. The present study aimed at filling this data gap by combining an expert survey on use of copper in individual crops with land-use data in 12 European countries, and therefore covering 84% of the European horticultural area. Other studies assessing copper use in organic agriculture have either focused on certain regions (e.g., Switzerland or Germany) or on specific crops (e.g., olives, citrus, tomatoes, and potatoes). The present survey takes into account reductions of copper limits (from 6 kg ha^{-1} y^{-1} to 28 kg ha^{-1} in 7 years) after 2018 as a result of regulation (EU)1981/2018, while other surveys which were conducted before 2018 use the limits in force at that time, which has to be taken into account when comparing results.

Copper Use Authorization and Estimated Real Use Per Hectare and Year

The extraordinary wide spectrum of copper is reflected in the presence of more than 60 permitted and more than 40 real uses identified in this survey. The wide range of activity emphasizes the particular challenge for substitution of copper-based fungicides, requiring a broad range of solutions tailored to each crop and plant pathogen. The number of allowed and real uses strongly varies between countries mainly as a result of the size of a country and the proportion of the organically managed area.

Furthermore, it also relates to the climatic conditions, i.e., whether weather conditions are favorable to grow certain crops or to occurrence and severity of certain plant diseases. Therefore, countries such as Spain, Italy, and France, by far, have the highest number of permitted and real uses.

Permitted amounts of copper ranged from 1.2 to 4 kg ha^{-1}, with a focus on 3 to 4 kg ha^{-1}. In more than 50% of these cases (crop × country), experts assessed that farmers use, on average, less than half of the permitted copper quantities. These results are consistent with results from other surveys. For example, in a Swiss survey among organic farmers performed in 2009–2012, utilization rates below 50% were found in nine out of 15 allowed uses. Low utilization rates in many crops reflects the high awareness of individual farmers to reduce the amount of copper, as well as the efforts of farmers' associations and national copper-reduction strategies. High utilization rates in particular crops in most surveyed countries can be indicative for crop protection issues which are difficult to solve without copper use, such as for grapevine and downy mildew (*Plasmopara viticola*).

In other cases, high utilization rates mirror restrictive copper limits as, for example, for Belgium, where the permitted amount of copper is set to 2–2.6 kg ha^{-1} (exceptions: potatoes, 4 kg ha^{-1}), resulting in utilization rates above 75% in most of the cases. In some crops, a high variability in the copper utilization rate between countries was observed. For example, in potatoes, on average, between 10 and 100% of the maximal legally permitted amount is used depending on the country.

This might reflect different climatic conditions and, thus, disease pressure: *Phytophthora infestans*, causing late blight in potato, prefers humid, cool conditions, and, consequently, utilization rates are high in Belgium, Germany, Switzerland, and the UK. For potatoes and tomatoes in Italy, the copper-use estimates in this survey are consistent, considering that, in Italy, limits have been reduced to 4 kg ha^{-1} y^{-1} in the meanwhile. In some other cases, the two studies' estimates of real copper use differ from each other, often with higher estimates in Katsoulas et al.

This discrepancy might reflect increased reduction efforts as a result of the reduced legally permitted maximum copper quantity after 2018 ((EU) 2018/1981). Furthermore, variability of copper uses depending on varieties or production systems could lead to different assessments by experts. For example, in olives, researchers found higher amounts of copper used in irrigated vs. non-irrigated systems, and in susceptible vs. less susceptible varieties, and Researchers report much lower copper utilization rates in downy mildew resistant ("PiWi") than in susceptible cultivars (13% vs. 75%).

Annual Copper Use

In view of the intended copper replacement in the coming years, an estimate of the quantities of plant-protection products needed for replacement is crucial. This survey estimates that more than 3200 t of copper needs to be substituted annually in the 12 surveyed countries to cover the needs of the current organic production area. Thus, with an average amount of 0.5 kg copper per ha and treatment, alternative products sufficient to treat 64,000 km^2 once a year would be necessary for the 12 surveyed countries. This calculation does not consider the foreseen increase in organically managed area or the potential increase of diseases due to climate change or the introduction of new plant pathogens. Due to the substantial lack of statistics on active substance use, the verification of copper-use data is not straightforward. At least for Germany and Switzerland, plausibility checks can be performed. In Switzerland, the sales of copper as a plant-protection product are 50 t y^{-1}.

This survey's estimate of 7 t y^{-1} used for plant protection in Swiss organic farming thus corresponds to approximately %14 of total copper use, which, in turn, is proportional to the organic farming area (i.e., %14 organic area share of total farmland in 2017, https://statistics.fibl.org/europe/area.html) (accessed on 7 March 2022), and thus plausible. For Germany, estimated a total use of copper of 26 t in organic farming in 2013. The 42 t y^{-1} calculated in this study (corresponding to an increase of %61) is in accordance with these findings, given the growth of the organic sector (%75+ in horticultural area from 2013 to 2017, statistics.fibl.org) and the partial compensation by successful copper-reduction strategies.

Not surprisingly, as for number of authorized uses, copper use in individual countries correlates with the organic area in which copper use is allowed, which obviously, in turn, is affected by the size of the country, the share of organically managed land, the cultivated crops, and the copper authorizations in these crops. In some countries (e.g., Germany and Switzerland), considerable efforts to reduce copper inputs at national levels have resulted in a significant reduction of copper use per ha in organic farms in the past decade.

In these countries, there is a common understanding that, with the current technology, no substantial further reduction is likely in the main crops, particularly in view of the increasing difficulties due to climate change and invasive diseases. In other countries where systematic reduction strategies have been initiated later than in Germany (e.g., Italy, Spain, and France), our experts and other studies assume that the reduction potential might not yet be fully exploited, especially in crops such as olives and nuts.

This study shows that, at a European scale, olives, grapevine, and nuts are the main drivers of overall copper use in organic farming, while crops such as apples, pears, and potatoes play a rather minor role. A plausibility check using estimates indicates that our estimates for olives presented here are rather conservative and that real copper use in olives might even be higher, with a significant impact on overall European copper use. While grapevine is widely known as a main copper-consuming crop, the importance of olives and nuts has not been well documented so far and was quite unexpected for many experts. This might result from the fact that these crops can only be cultivated in relatively few Mediterranean countries, and from the rapid growth of the organic production area in these countries. For example, in the countries studied, organically managed area for olives and nuts increased 2.4-fold and 7.8–fold, respectively, between 2004 and 2017. Other crops, such as potatoes, which are generally considered to be very copper-intensive, contributed little to total copper consumption, probably due to the relatively small area under cultivation (e.g., potatoes equal 0.65% of the total organically farmed area in the countries studied) and considerable efforts to reduce copper use. Nevertheless, these crops may be important for copper consumption in certain countries, e.g., for potatoes in Belgium, Germany, Switzerland, and the United Kingdom.

Copper-Replacement Strategies

In view of the considerable amounts of copper to be replaced, it is evident that only a consequent combination of all preventive measures and new alternative products can be successful. The consequent exploitation of preventive measures is a core principle of organic farming that is reflected in the EU regulations, the guidelines of FAO/WHO, and the private standards of IFOAM and farmer associations. For example, the use of disease-resistant cultivars and, more recently, the use of rain shelters in horticultural crops are efficient component strategies for copper reduction that are exploited to different degrees depending on crops/countries. The availability of resistant cultivars is a key factor in helping copper reduction.

For example, large differences in copper dependency were described for different olive varieties having different levels of susceptibility to diseases, and considerably lower amounts of copper are used in more resistant grapevine (PiWi) varieties. For example, Swiss organic farmers, on average, only use 13% of the maximally allowed copper amounts in PiWi varieties, in comparison to 73% in traditional susceptible varieties. In the Netherlands, where the use of copper as a plant-protection product is not allowed, the yield of organic potato per ha could be increased by 80% by introducing new resistant potato varieties. In many countries, efforts are currently being made to open up markets for new resistant/tolerant varieties. However, preferences of consumers and/or retailers concerning variety and related quality attributes often dictate what farmers should produce.

Consequently, the potential market of resistant/tolerant varieties has not yet been fully exploited in many crops/countries. Overall, a successful market introduction of new varieties requires strategies that are tailored to the specific context encompassing the entire supply chain from breeder to consumer. The adoption of preventive strategies (e.g., planting of robust varieties in perennial crops and use of rain shelters) is often costly, time-consuming, and risky for farmers if the market is not receptive. Furthermore, changes of production systems may trigger the emergence of currently minor or secondary diseases or the development of virulent pathogen strains that may overcome varietal resistance.

The first phenomenon was observed when apple scab-resistant cultivars were planted at scale: the consequent reduction of fungicides against the main pathogen let previously irrelevant diseases, such as *Diplocarpon mali*, become important. In the second case, for example, new pathogenic races of apple scab–overcoming resistance genes were continuously emerging.

Even if all preventive crop-protection strategies are implemented, there is still a significant need for alternative plant-protection products to substitute copper. There are several alternative products available, but none can compete with copper in terms of spectrum of activity, efficacy, and price for growers. While some products with acceptable-to-good efficacy are available for the control of ascomycetes and basidiomycetes (e.g., sulfur, aluminum sulfate, potassium hydrogen carbonate, and lime sulfur), the control of oomycetes and bacteria is provided only to a limited extent by some products. Therefore, crop protection in plants with a wide range of bacterial diseases (e.g., hazelnut, walnut, and oranges) still heavily depends on copper-based plant-protection products. However, the national experts of this survey assess that the potential of available alternatives is probably not fully exploited in some crops, such as olives or nuts, which are often grown with traditional methods, where comparably little effort has been invested into the development of novel strategies. Therefore, efforts to reduce copper use in these crops should be intensified.

The urgent need for highly active fungicidal compounds of natural origin has been recognized, and systematic screenings for novel compounds were initiated by several research groups, as well as the industry. Promising new plant-protection product candidates include an extract from *Larix decidua*, the monosaccharide tagatose derived from lactose; an extract from liquorice (*Glycyrrhiza glabra*); and a pelargonic acid–based product.

Furthermore, refined inorganic compounds, such as calcium carbonate or calcium hydroxide, have been further developed for this aim. If the alternatives are botanical extracts, upscaling to the staggering quantities needed will be a major challenge, while micro-organisms are generally more readily up-scaled in large fermenters. The evaluation and further development of alternatives represented an integral part of several EU-funded projects (e.g., RELACS (https://relacs-project.eu/), FORESTSPECS. Even though the most advanced compounds show promising levels of efficacy and seem to cover different uses, none of these alternatives has been authorized as a plant-protection product so far in the EU. The EU has a long and complex procedure to authorize the placing on the market of novel plant-protection products, which requires substantial efforts and investments.

This process guarantees citizens and consumers that plant-protection products do not pose substantial risks for health and environment. However, current procedures for bioprotection products (including biocontrol organisms and botanicals) in the EU have evolved from regulations for synthetic active substances. The current lack of specialist bioprotection regulatory body, legislation, procedure, and data requirements is made responsible for the low number of plant-protection products of natural origin placed on the market. The International Biocontrol Manufacturers Association (IBMA) therefore proposes that the EU, for example, establish a bioprotection-specific body and encourages the development of tailored data requirements to facilitate market introduction of products of natural origin, which are important tools to reduce copper use and thus to contribute to the implementation of the Sustainable Use of Pesticides Directive of the European Union.

FUNGICIDE USE IN FIELD CROPS: BENEFITS AND RISKS

Plant diseases are caused by many different organisms including fungi, bacteria, viruses, nematodes, and other organisms. Fungicides are pesticides used for controlling fungal and fungal-like diseases and can be applied as a seed treatment, in-furrow application or foliar spray. They include both synthetic and non-synthetic options such as copper, sulfur and oils.

There are also biological control options and products that induce the immune system of the plant, but here we will focus on synthetic fungicides. Fungicides work, in general, by blocking a specific metabolic pathway in the fungus that prevents spore germination or hyphal growth. These different mechanisms are called "modes of action" (MOA).

Plant diseases can be difficult to diagnose. Diagnosis is easier with experience, but comes down to a combination of familiarity with symptoms and look-alike symptoms, an investigation of the pattern and timing of symptom appearance, and, when needed, testing for the pathogen of interest for confirmation. When you need assistance with disease identification, contact your local county or reservation Extension agent or the Schutter Diagnostic Laboratory (406-994-5150).

Treatment Options

After you have identified the plant disease you are concerned with and confirm that it is fungal, treatment options can include fungicides. However, not all fungicides are effective against all fungal diseases. Good sources of information about fungicide efficacy and rates include the product label and Extension sources. We have tables that list the products that are currently registered and their efficacy or registration status on some widely planted crops in Montana such as wheat, pulses, and potatoes.

Fungicides for Disease Management

When using fungicides for disease management, the principles of Integrated Pest Management (IPM) should be used to avoid resistance development, including:

Preventative Cultural Practices

Use best management practices including using high quality, pathogen-free seed, crop rotation, using an adapted crop variety, optimal seeding rate, planting date, irrigation practices, fertilization, sanitation including breaking the 'green bridge,' etc.

Monitoring

Scout your crop for pests regularly and get pests accurately identified. Use degree-day models where available to determine when the pest is likely to reach medium-high to high risk.

Acceptable Pest Levels

Determine what level of the pest you are willing to tolerate.

Mechanical Controls

Remove infected plants from the system to prevent reproduction and spread of the pathogen (rogueing).

Biological Controls

Natural biological systems can mitigate pest damage. Beneficial insects that predate on or parasitize insect vectors of plant viruses and biological controls.

Responsible Use

When a pesticide is needed, follow all label restrictions and use the best application methods possible to target the disease of interest. If lack of efficacy is suspected, leave an untreated strip to compare with treated areas.

If level of disease is the same, then a symptomatic sample should be sent to a diagnostic clinic.

Other Recommendations

Practices to prevent fungicide resistance:

- Select and use fungicides correctly.
- Rotate the use of fungicide MOA.
- Limit number of applications of fungicides in a particular MOA each season, including seed treatment.
- Mix MOA in blends or tank mixes.
- Use fungicides at recommended rates.
- Follow all label directions.

8

USING FUNGICIDES TO MANAGE DISEASES ON FIELD CROPS

Planting resistant cultivars is one of the least expensive and most effective management strategies to prevent diseases. However, cultivars with an adequate genetic resistance level to necrotroph foliar diseases are scarce, and usually resistance to leaf rust is complete, conditioned by one or a few genes and has low level of durability in Argentina. Therefore, chemical protection together with cultural practices is a common method of control. In addition, fungicides are also important because Argentinean wheat region combine high yield potential cultivars with high infection pressure, both deriving from adequate temperature and moisture levels, large application of N fertilizers and rotations dominated by cereals, which promote progression of some foliar diseases.

However, the response varies depending not only on the fungicide but also on the N fertilization level, tillage system, foliar disease type and characteristics of the genotypes. The relationship between yield loss and disease severity can differ widely between crop genotypes and some of them exhibit a smaller yield loss under a given severity of infection than others. On the other hand, mechanisms of fungicides to control foliar diseases on wheat may vary according to the active ingredient they have.

Recently, varieties with French germplasm have been introduced or crossed with local germplasm to produce new cultivars in Argentina. These cultivars are characterized by high yield potential but lower resistance to foliar pathogens as tan spot, leaf blotch and leaf rust than the traditional ones. However the increasing adoption by growers of French germplasm varieties susceptible to foliar diseases is leading to a higher use of fungicides.

Triazoles and Strobilurins are the most common systemic fungicides used to control foliar diseases on wheat in Argentina. Statistics shown by Campos indicate that 50% of the products used in Argentina are triazoles and the remaining 50% consists in mixtures of formulations containing triazoles and strobilurins. Systemic fungicides are absorbed through the foliage or roots and are translocated within the plant through the xylem.

These types of fungicides generally move upward in the transpiration stream and may accumulate at the leaf margins.

Triazoles are characterized by being an active ergosterol inhibitor, which is the major sterol in fungi. Sterols derivate from terpenes, and they are an essential part of the fungal cell membrane. These molecules are rigid and flat and in its association with the cell membrane give them stability, making it less flexible and allowing the permeability control. Ergosterol Biosynthesis Inhibitors (EBIs) have become one of the most important groups of fungicides, however they may not be effective in controlling Oomycetes because they do not possess the ergosterol synthesis.

The EBIs can be divided into: 1,4 α–demethylase inhibitors (DMIs), which includes the azole (triazole, imidazole) and pyrimidines; the Δ 8,7 isomerase and Δ 1,4 reductase inhibitors (morphines and piperazines) and 3-ceto reductase sterol inhibitors (hydroxyanilide).

The triazoles have been useful to control many foliar diseases. They inhibit the fungus dependent enzyme cytochrome P-450 called 1,4á-demethylase involved in the ergosterol biosynthesis and consequently affect the permeability of the membrane. However, the mode of action may vary relatively between the different active principles within this group. One of the most common chemicals commercialized in Argentina containing triazoles is Tebuconazole, which is used for seed treatment and foliar and spike applications in cereals.

The fungi-resistance genetic basis to triazoles is not well known. In many cases it seems to be polygenic and observed decreasing effectiveness does not always imply loss of yield performance. The triazol group has many benefits such as high antifungal activity, low toxicity to other organisms, curative properties, and they are compatible with an integrated disease management; however its preventive action is low. That is why they are usually used in mixed formulations with other chemical groups to compensate this deficiency.

THE USE OF FUNGICIDES IN THE INTEGRATED FOLIAR DISEASE MANAGEMENT TO ENHANCE WHEAT YIELD AND QUALITY

Crop potential yield is defined as that attainable yield, when no nutrient or water limitations occur, i.e. when incident radiation, temperature and physiological crop genotype characteristics determine yield. On the other hand, grain quality has several definitions depending on the users; therefore, the end-use quality is vastly diverse. Several factors have influence on the severity of the main foliar diseases of wheat, among them resistance of the cultivars, tillage systems, N fertilization and fungicide applications.

Genetic resistance is the basis of the integrated disease management. Plant disease resistance can be classified into two categories: qualitative resistance, conferred by a single resistance gene (also termed as race non- specific or slow rusting resistance) and quantitative resistance, mediated by multiple genes or quantitative trait loci (QTLs) (also termed as race non-specific or slow rusting resistance) with each providing a partial increase in resistance.

Considering the main foliar diseases in wheat during the last decade, 18 major genes conferring resistance to the pathogen have been identified for resistance to *Septoria tritici*. They were: *Stb1* located on the chromosome 5BL, *Stb2* on the chromosome 3BS, *Stb3* on the chromosome 6DS, *Stb4* on the chromosome 7DS; *Stb5* on the chromosome 7DS; *Stb6* on the chromosome 3AS; *Stb7* on the chromosome 4AL; *Stb8* on the chromosome 7BL; *Stb 9* on the chromosome 2B, *Stb10* on the chromosome 1D. *Stb 11* on the chromosome 1BS, *Stb12*, on the chromosome 4AL,*Stb13* on the chromosome 7BL, *Stb14* on the chromosome 3BS,*Stb15* on the chromosome 6AS, *Stb16* on the chromosome 3D, *Stb17* on the chromosome 5A and *Stb 18* on the chromosome 6DS. In addition, several QTL were also found. Eriksen *et al.* found some on chromosomes 2BL, 3AS, 3BL, 6B and 7B. In Argentina resistance was localized in several foreign lines

Considering resistance to tan spot eight races of the pathogen has been characterized based on their ability to cause necrosis and/or chlorosis in differential wheat lines. In Argentina and in general around the world cultivars with acceptable levels of resistance to tan spot and Septoria leaf blotch are scarce.

Considering leaf rust, more than sixty genes for leaf rust resistance (*Lr*), most of them major or race specific genes, have been catalogued to date in wheat. However, the gene-for-gene interaction between host resistance genes and pathogen virulence genes combined by virulence shifts in pathogen populations have reduced the effectiveness of a significant number of major leaf rust resistance genes. Replacement of highly variable land races by higher yielding, pure-line varieties in many parts of the world, including the South Cone, has further reduced the wheat gene pool and favored virulence shifts events in pathogen populations.

In Argentina using molecular markers, a set of 66 adapted cultivars previously evaluated by gene postulation for presence of 15 *Lr* genes was screened, and eight genes were detected: six seedling genes (*Lr9, Lr10, Lr19, Lr24, Lr26, Lr47*) and two adult plant resistance genes (*Lr34, Lr37*). Genes *Lr20, Lr21, Lr25, Lr29, Lr35* (adult plant resistance gene) and *Lr51* were not detected in tested cultivars. Resistance in most Argentinean cultivar and around the world is conditioned by one or a few genes.

In the Rolling Pampa region of Argentina, conservation management practices such as no tillage are increasing as alternative cropping systems. No tillage systems have been implemented to restore soil structure in large areas cultivated with double-crop sequences such as wheat (*Triticum aestivum* L.)/soybean (*Glycine max* L. (Merr.); corn (*Zea mays* L.) -wheat/soybean; or wheat monoculture. Annual wheat/soybean double-crop sequences using conventional tillage are considered less desirable because of the effect on soil organic matter and the reduced quantity of residue that soybean crops leave after-harvest. In the semiarid region of Argentina, conservation management techniques are also necessary to prevent soil erosion and effectively store and use the limited amount of precipitation for crop production. No tillage can also reduce costs by decreasing fuel consumption required to produce a crop.

However, in the wheat/soybean system under no tillage, as in wheat following wheat, the inoculum of necrotrophic fungi usually survives until the next wheat season; typically, a minimum of one to two years between wheat crops is required to reduce populations of these organisms. In no tillage systems, crop residue mineralization is slow. It requires 14 to 16 months in Brazil but approximately 18 to 32 months in Argentina and Uruguay due to lower average temperatures than in Brazil. No tillage may have a different effect on plant diseases depending on the soil type, geographic location, environment, and the biology of the particular disease-causing organism.

Tan spot and Stagonospora blotch[*Phaeosphaeria avenaria* (G.F. Weber) O. Eriksson f. sp. *triticea* T. Johnson, anamorph *Stagonospora avenae* (A. B. Frank) Bissett f. sp. *tritica* T. Johnson] increased in no tillage systems in wheat monoculture or wheat following fallow, although the opposite occurred when wheat followed other crops. In some studies, conventional tillage increases crop residue mineralization, reducing fungal inoculum. However, others reported contrasting results regarding the effect of no tillage on necrotrophic wheat diseases, depending on the environment and the crop growth stage evaluated (early or late in the season).

Fungicides are widely used to manage foliar wheat diseases in Argentina and several countries. The response to fungicide application depends on the severity of specific foliar diseases, cultivar disease resistance or tolerance, management practices, and environmental conditions. Fungicides applied at flag leaf and spike emergence of winter wheat increased mean grain weight and grain yield when they extended canopy life . The green area duration of flag leaf is important because is the last leaf senescing, it intercepts more light than lower leaves and it is in closer vascular proximity to spikes than lower leaves.

Strategies to protect flag leaf and delay the senescence process are therefore important to assure not only higher yield but also higher grain quality. Goodingfound that the effect of fungicides increasing green area duration of the flag leaf was associated with increases in yield, thousand grain weight and specific weight. Fungicides containing strobilurins to control foliar diseases in wheat are associated in some cases with higher increases in grain yield and grain weight comparing with triazoles. Dimmock & Gooding reported that strobilurins prolonged green flag leaf area duration and increased mean grain weight significantly more than triazoles.

Researchers reported wheat yield increases following fungicide treatments ranging from 0.8 to 4.4 Mg ha^{-1}, depending on the amount of infested straw on the soil surface, disease severity and fungicide strategy (type of active ingredient, timing or number or applications, rates and method of application). Severe foliar infections before or at flowering stage of wheat are extremely damaging and may cause important yield losses, whereas when serious infections occur later, the damage to yield is much smaller.

Increased yields disease management are associated mainly with an increase in thousand grain weight, while other yield components such as number of spikes.m^{-2} or grains.$spike^{-1}$ are usually not affected by disease severity. However, Simón *et al.* reported that preventing early wheat infection by *Septoria tritici* could result in an increase of spikes.m^{-2} and grains.$spike^{-1}$.

In Argentina, Serrago *et al.* determined that grain number was not affected by foliar diseases when they appeared after anthesis. Grain weight was strongly, poorly or not affected by foliar diseases and was not associated individually with both, the sink size and the source size. However, when the grain weight increment due to fungicide application was plotted against the healthy area absorption per grain, a significant negative association was found for the Argentine experiments.

When the healthy absorption area per grain was corrected by the grain weight potential all experiments conducted in Argentine and in France fit well to a common negative linear regression for the relationship between grain weight variation and grain weight potential demonstrating that grain weight potential is an important feature to consider in diseases control programs.

Foliar diseases forced the crop to use the accumulated reserves increasing the utilization rate of the water soluble carbohydrates, depleting as a consequence the water soluble content at physiological maturity in all experiments. The association between water soluble carbohydrates and the healthy area absorption per grain corrected by grain weight of healthy crops suggests that foliar diseases in wheat cause source limitation, forcing to the crop to use the water soluble content reserve which could be insufficient to fill the grains previously formed.

Management practices such as N fertilization can also affect the expression of wheat foliar diseases and the effectiveness of foliar fungicide application. Increasing N rates may cause negative, positive or neutral effecton foliar disease severity, depending on the geographic location and the type of disease. The magnitude and direction of the influence of N supply on Septoria leaf blotch severity has been studied with contrasting results. Simón *et al.* found that in conducive conditions, N fertilization increases the severity of Septoria leaf blotch and discussed the effect of different factors affecting the influence of N supply. Increasing N rates retarded tan spot development.

However, Bockus and Davis suggested that N applications do not directly affect tan spot severity, but rather appear to reduce disease impact through delayed leaf senescence or that high N rates increase Septoria leaf blotch or tan spot severity due to an increase in crop biomass production, which creates a micro-environment conducive to fungal development in humid regions. In addition, experiments carried out in Argentina indicated that yield increase and increase in yield components due to application of tebuconazole was similar in fertilized and non fertilized conditions, despite the increase in the area under disease progress curve under N fertilization.

Biotrophic pathogen such as leaf rust also causes important diseases in wheat. N fertilization usually increases the severity of this disease.

CORN

Introduction

Corn, (*Zea mays*), also called Indian corn or maize, cereal plant of the grass family (Poaceae) and its edible grain. The domesticated crop originated in the Americas and is one of the most widely distributed of the world's food crops. Corn is used as livestock feed, as human food, as biofuel, and as raw material in industry.

Corn was first domesticated in southern Mexico about 9000 years ago. Its closest wild relative is teosinte, a wild grass of Mexico, Guatemala and Honduras. A major puzzle is the great genetic differences between teosinte and corn, indicating how key mutations and human selection contributed to genetic evolution. After the Columbian exchange, corn production spread throughout the world. Corn is a highly productive crop with the ability to exploit available soil nutrients. As a C4 plant, corn has some photosynthetic advantages in capturing solar energy in warm weather compared to C3 crops such as wheat, rice and soybean. Due to its high productivity under various climate conditions, corn is now the largest grain crop in the world. Favorable agro-climatic conditions in the US "Corn Belt" have made the US the largest corn producer.

The rise of corn as the most important cereal in the world has been associated with important improvements in its productivity. Starting in 1940, a period of fast and steady rise in corn productivity began and continues to this time. US average corn yield increased from 28.9 bu/acre (1.81 metric tons/ha) in 1940 to 176.6 bu/acre (11.1 metric tons/ha) in 2017. This amazing achievement means that a given area of land can produce 6.1 times more corn in 2017 than in 1940, which corresponds to an average annual growth rate of 2.35%, reflecting the rapid technological progress sustained over the last seven decades. This achievement raises two questions. First, what are the sources of this growth in corn productivity?

We discuss the role played by two key drivers of corn productivity: improved genetics and improved management. We also consider the corn market and its evolving prices. Finally, we reflect on what may come next. Some evidence suggests that agricultural productivity growth may be slowing down, raising concerns about our ability to feed a growing world population. We ponder these prospects as they apply to corn production.

Fungicides are an effective in-season management tool for fungal leaf diseases, and sometimes can reduce chances of stalk rot development. A number of fungicide products that are effective against fungal pathogens on corn are available for use. However, some are more effective on certain pathogens than others. Most of the fungicides available are preventive in nature and stop the fungus from infecting or advancing within the plant. Therefore, timing of a fungicide treatment is critical. If fungicides are applied when the severity is already high, the benefit will be limited.

Physical description

The corn plant is a tall annual grass with a stout, erect, solid stem. The large narrow leaves have wavy margins and are spaced alternately on opposite sides of the stem. Staminate (male) flowers are borne on the tassel terminating the main axis of the stem. The pistillate (female) inflorescences, which mature to become the edible ears, are spikes with a thickened axis, bearing paired spikelets in longitudinal rows; each row of paired spikelets normally produces two rows of grain. Varieties of yellow and white corn are the most popular as food, though there are varieties with red, blue, pink, and black kernels, often banded, spotted, or striped. Each ear is enclosed by modified leaves called shucks or husks.

Types of corn

Commercial classifications, based mainly on kernel texture, include dent corn, flint corn, flour corn, sweet corn, and popcorn. Dent corn, primarily grown as animal feed and for food manufacturing, is characterized by a depression in the crown of the kernel caused by unequal drying of the hard and soft starch making up the kernel. Flint corn, containing little soft starch, has no depression; it is used for decoration and is eaten as hominy in the Americas. Flour corn, composed largely of soft starch, has soft, mealy, easily ground kernels and is an important source of corn flour. Sweet corn, commonly sold fresh, frozen, or canned as a vegetable, has wrinkled translucent seeds; the plant sugar is not converted to starch as in other types. Popcorn, an extreme type of flint corn characterized by small hard kernels, is devoid of soft starch, and heating causes the moisture in the cells to expand, making the kernels explode. Improvements in corn have resulted from hybridization, based on crossbreeding of superior inbred strains.

Genetically Modified Corn

Many industrial and feedstock varieties of corn are genetically modified organisms (GMOs) engineered for resistance to the herbicide glyphosate or to produce proteins from *Bacillus thuringiensis* (Bt) to kill specific insect pests. In addition, some strains have been genetically engineered for greater drought tolerance and to increase their nutritional value. Most of the corn grown in the United States is GMO, which may reduce the need for herbicides and insecticides.

Plant breeding was once restricted to sexually compatible plants, and generations of offspring were selectively bred to create unique varieties. In fact, corn, along with rice and wheat—today's global crop staples—would not exist without such techniques. With the goal of ever-widening the pool of genetic diversity, conventional plant breeding has gotten more technologically savvy in recent years. For example, realizing that natural mutants often introduce valuable traits, scientists turned to chemicals and irradiation to speed the creation of mutants.

From test-tube plants derived from sexually incompatible crosses to the use of molecular genetic markers to identify interesting hereditary traits, the divide between engineering and genetics was narrowing long before kingdom boundaries were crossed.

But when geneticists began to explore microorganisms for traits of interest—such as *Bacillus thuringiensis* (Bt) genes that produce a protein lethal to some crop pests—they triggered an uproar over ethical, scientific, and environmental concerns that continues today.

Uses and products

Corn is one of the world's most productive and dominant crops. It is grown extensively as food for both humans and livestock, as a biofuel, and as a crude material in industry.

Food and nutrition

Corn is the third largest plant-based food source in the world. Despite its importance as a major food in many parts of the world, corn is inferior to other cereals in nutritional value. Its protein is of poor quality, and it is deficient in niacin. Diets in which it predominates often result in pellagra (niacin-deficiency disease). Corn is high in dietary fibre and rich in antioxidants.

Unlike many other cereal grains, corn flour is gluten-free and cannot be used alone to make rising breads. It is widely used, however, in Latin American cuisine to make masa, a kind of dough used in such staple foods as tortillas, arepas, and tamales.

In the United States and many other places, sweet corn is boiled or roasted on the cob, creamed, converted into hominy (hulled kernels) or meal, and cooked in corn puddings, mush, polenta, griddle cakes, cornbread, and scrapple. It is also used for popcorn, confections, and various manufactured breakfast cereal preparations.

Proactive Fungicide Treatments

Economic Benefit

Several research studies have shown that when a fungicide is applied in the absence of disease or very low disease severity, the probability of increasing yield to pay for the treatment decreases significantly. For example, researchers analyzed data from 613 treatment comparisons of strobilurin-treated and nontreated plots over a 10-year period in the Corn Belt region. The fungicides were applied between V14 and R5 (dent) with a majority of treatments being applied between tasseling (VT) and R2 (blister).

They reported that when disease severity was less than 5% on the ear leaf at the end grain fill period, the fungicide treatment increased yields 1.5 bu/acre, and when the disease severity was > 5% the yield gain averaged 9.6 bu/ acre. These results suggest that there may be some benefit from proactive fungicide applications. The yield enhancement has been linked to improved crop health (stays green longer) and reduced fungal populations. However, these benefits must be balanced against the long-term risk of the fungal pathogens developing resistance. Therefore, to avoid problems associated with unnecessary application of fungicides (such as resistance development, added expenses), growers should always scout to determine the need for a fungicide application.

Fungicide Use and Activity

Fungicides used for in-season corn and soybean disease management are primarily in the chemical group of either strobilurin or triazole fungicides. Also, there are many pre-mix products that are available that contain both strobilurins and triazoles. Strobilurin and triazole fungicides are considered "locally systemic" which means they are absorbed into plant tissue. However, there is minimal systemic activity. For example, the fungicide may move slightly outward through a leaf from the point of uptake but does not move downward through the leaf stem, therefore new plant growth after application will not be protected.

Fungicides have preventative and/or curative activity. If a fungicide is on the plant before infection occurs it may act as a protective barrier preventing infection. Curative activity may occur when the fungicide is present within plant tissue and stops early growth of the pathogen in plant tissues. Fungicides with curative activity will not "cure" a plant from disease and are most effective if applied prior to infection or within the first 72 hours after infection.

Effective fungicide performance requires good spray coverage. After application, fungicides degrade from the leaf surface and are generally considered effective for 21 days. Also, foliar-applied fungicides are ineffective in protecting plants from fungal pathogens that primarily enter through plant roots and cause diseases such as seedling diseases and crown rot in corn; also soybean diseases such as phytophthora root rot and sudden death syndrome.

Assessing Disease Risk

Factors contributing to higher risk of disease for corn and soybean fields are similar and include the following: susceptible seed product, continuous corn or continuous soybean systems, history of disease infection in the field, current disease activity in the field, current and projected disease-favorable weather, high plant population/dense crop canopy, high yield potential, irrigation, and no-till or reduced tillage systems. Assessing individual fields for their risk of disease can serve as aid in identifying which fields should be prioritized for scouting.

Foliar fungal diseases of primary concern in Michigan are northern corn leaf blight and gray leaf spot. The fungal pathogens that cause these diseases survive on residue, so fields under minimum tillage and corn-on-corn rotations are at greatest risk. Spores from both of these pathogens can either be splashed up onto foliage or deposited by the wind.

Northern corn leaf blight can be identified by the distinctive lesions that form on the leaves. Lesions start out as light green, but develop into the tell-tale tan or gray "cigar"-shaped lesions. Gray leaf spot lesions typically appear two to three weeks prior to tasseling as narrow, long, rectangular (up to 2 inches) lesions and are light tan, typically delineated by the veins.

Crop rotation will reduce the amount of inoculum of northern corn leaf blight and gray leaf spot present at the start of the season, and tillage can also help to break down corn residue. Scout fields before making a fungicide application. When scouting fields, walk into the field beyond the headland as disease is typically worse along the edge of the field, especially since that is where it is exposed to prevailing winds.

Fig. 1 Gray leaf spot on corn

Common rust is another disease typically seen, but often not at economic levels. The rust fungi require a living host to survive. Urediniospores of common rust survive the winter on corn in the southern United States, and are then carried long distances by wind to reach the Midwest. Rust pustules typically appear in late June and are favored by high humidity and moderate temperatures. Most hybrids are fairly resistant and common rust severity is typically not severe enough to warrant selecting resistant hybrids. Crop rotation does not influence common rust disease as it does not survive on residue.

Foliar Fungicide Application

A rather new practice in the Northeast is the application of foliar fungicides at late vegetative stages through kernel blister stage (R2) to control foliar fungal diseases, in particular gray leaf spot and northern leaf blight, on the upper leaves during grain filling. The primary fungicides being applied in New York are products that combine a strobilurin and a triazole fungicide. The cost of product plus application is currently in the range of $35 to $45 per acre.

There are no firm disease thresholds to guide fungicide application at this time, though foliar fungicide application is generally indicated in situations where fungal diseases have developed in a majority of plants at or just below the ear leaf by the time of tassel emergence (VT). The risk of foliar diseases (and therefore the potential benefit of fungicide) is increased in susceptible hybrids, in continuous corn under reduced tillage, where there is a previous history of disease, and in humid environments with persistent morning dew.

Since economic return on foliar fungicide investment is still poorly understood, growers who apply fungicide are urged to leave non-sprayed strips in their fields for comparison of yield and disease severity. The below table lists labeled fungicides based on their relative efficacy in controlling foliar diseases that occur in New York. Follow label directions carefully for all fungicide applications. Indiscriminate application of foliar fungicides to all corn acres would likely hasten the development of pathogen strains that are resistant to these fungicides and is therefore discouraged.

Fungicide Resistance

Fungicide resistance is when a fungicide that used to control a given fungal pathogen, no longer offers any protection against the same fungus. Several factors are responsible for fungicide resistance including:

- The fungicide provides a selection process for pathogens that are resistant or tolerant to the treatment.

  Practices that increase the risk of fungicide resistance include:

  - Multiple applications of fungicides with same mode of action.
  - Reducing the application rate or using off-label products.
  - Multiple applications of the same mode of action fungicides within a single year.

- High genetic variability within the pathogens.
  - High variability suggests that some pathogens will have inherent tolerance to fungicide.

- High reproduction capacity of the pathogen.
    - Pathogens, which reproduce quickly (e.g., rusts), are likely to have increased diversity and therefore likely to be selected for when fungicides of similar modes of action are applied to the same area in a season.

To avoid fungicide resistance, growers should monitor the performance of the fungicides they use. One way to do this is to leave a strip of a nontreated area (one pass), where the yield and disease severity from the treated and nontreated zones can be compared. If the two areas have comparable disease severity, this would mean that the fungicide had minimal impact on the disease that year and this could probably be due to resistance development. In this case, samples of diseased leaves in the treated area and untreated area should be collected and sent to the SDSU Plant Diagnostic Clinic for fungicide sensitivity testing. Proper disease identification and appropriate fungicide selection is crucial for effective use of fungicides.

The risk of developing fungicide resistance can be reduced by:

- Rotating between different classes of fungicide within a season and also between seasons.
- Scouting to determine the need for a fungicide and avoid applying fungicide when it is not necessary or when it is too late (severe symptom on ear leaf and higher).
- Using a mixture of fungicide classes. Luckily, several fungicide products are "broad spectrum."
- Practicing integrated disease management to reduce disease pressure.
- Following the label directions to determine the rates, growth stage of the crop, compatibility with other pesticides, and safety information.

Fungicide Classes

Fungicides are classified into groups depending on their mode of action. For instance, some fungicides interfere with fungal protein synthesis, while others interfere with respiration, etc. The Fungicide

Resistance Action Committee (FRAC) is an international committee that is responsible for fungicide resistance monitoring. The panel has developed FRAC codes that classify fungicides into classes with the same mode of action. Fungicide labels contain the FRAC code and fungicides with the same

Host Resistance in Management of Plant Diseases

Cultivar selection is a critical step in integrated pest management (IPM). Prior to the use of synthetic chemicals, farmers chose and saved seed from the best yielding and healthiest plants (e.g., bigger corn ears) for the next growing season. Today, planting a carefully selected corn hybrid may be the most important management decision to get maximum yield. Corn hybrids are developed to suit different needs: maturity, resistance to pests and diseases, plant characteristics (plant height, seed color, stalk strength, etc.).

Host resistance/tolerance, when available, is the first line of defense in plant disease management. Disease resistance genes have been bred into hybrids through conventional breeding or genetic engineering. Evolutionarily, plants and pathogens have co-existed together. When plants are attacked by a pathogen, the

Corn Disease Management: When to Apply a Fungicide?

The choice of whether or not to apply a fungicide will vary depending on several key questions. These are:

- What is your yield potential?
- What are the variety's resistance traits?
- What is the current growth stage of the corn?
- Is disease present in the field or locally?
- Was a fungicide already applied?

It is critical to assess how the potential yield savings from a fungicide application will affect your budget, and if resistant traits are present for the disease of interest. Budget tools, such as UGA's corn budget, can be very helpful in this decision process. Fungicides can usually save between 5 and 10 bushels per acre, but these values can vary depending on the amount of disease present in the field. Thus, it will be critical to evaluate a range of yield savings and the impacts they will have on the overall returns.

Growth stage is also important when considering the benefits of a fungicide application. An application in the V-stages can reduce disease severity in corn, but often does not lead to significant yield increases. In general, fungicide applications around VT (Tassel) to R2 (Blister) provide the best yield savings. There are situations where sprays outside these growth stages can be beneficial and further information about growth stage timing can be found at your local extension office.

The timing of the first fungicide application and deciding whether or not to apply a second application requires gathering some information about the disease. If the disease is present in the field or locally (e.g. neighboring county) and environmental conditions are conducive for disease development, a fungicide spray has a high probability of saving corn yields. Monitoring corn diseases should be primarily done through scouting, however, there are multiple resources available to help with corn disease monitoring. The decision of a second fungicide application depends on when the first application was applied, the growth stage of the crop and if conditions remain conducive for disease. Basically, it requires going through all the questions again, and assessing the impact and duration of protection from the first fungicide. About two weeks after the first application is a rough time estimate to begin considering a second application and evaluating the impacts of a fungicide spray.

Fungicide Research

Scientists, agronomists, and university collaborators have conducted several corn fungicide studies in which a single foliar disease was predominant at the research location or locations. In some cases, research locations were chosen specifically due to their history of a specific disease; in others, environmental conditions happened to be favorable for a given disease when the study was conducted.

Gray Leaf Spot

A research project was conducted over three years at the University of Tennessee Research and Education Center at Milan at a research site specifically chosen due to a history of high gray leaf spot pressure. The primary goal of this study was to determine the yield benefit associated with foliar fungicide management of gray leaf spot in hybrids with differing levels of genetic resistance. The plot area was in irrigated no-till corn production for four years prior to the start of the study, with a high level of gray leaf spot each year. Three Pioneer brand corn hybrids with differing levels of resistance to gray leaf spot were included in the study.

Results of the study demonstrated the potential for gray leaf spot to cause substantial reductions in yield when disease pressure is very high. Hybrid resistance was effective in mitigating a large portion of yield loss due to gray leaf spot; however, even with the most resistant hybrid, the yield benefit of the foliar fungicide application was great enough to potentially cover the cost of product and application. Under more moderate disease pressure, a fungicide application would likely not provide an economic benefit on a resistant hybrid.

Common Rust

Scientists conducted fungicide research trials at several Midwestern sites in 2009, a growing season that experienced unusually high common rust pressure in parts of the Midwest. Summer temperatures were cooler than normal in 2009, which favors development and spread of common rust. Studies were conducted at ten different field locations across five states. Corn yield response to fungicide application varied widely among research locations, largely due to differences in common rust pressure. Common rust was prevalent at research locations in Iowa, Illinois, and Indiana.

Common rust was prevalent at a trial at Macomb, IL, along with low to moderate levels of gray leaf spot and northern leaf blight. Notable differences in disease symptoms and yield response to fungicide were observed (Fig. 2).

These research results from 2009 demonstrate the value of foliar fungicides in protecting yield when disease outbreaks occur; however, genetic resistance of hybrids may also provide adequate protection and should be considered in fungicide treatment decisions.

Fig. 2. Two hybrids treated (left) and non-treated (right) with fungicide at Macomb, IL. The fungicide helped to protect yield of a susceptible hybrid (top) but provided little benefit on a moderately resistant hybrid (above).

Northern Corn Leaf Blight

Scientists on-farm trials were conducted at 40 locations in Iowa in 2015 to evaluate corn yield response to foliar fungicides applied at different timings. Northern corn leaf blight pressure was high in much of Iowa in 2015 and it was the predominant foliar disease at the trial locations. Trials compared yield of corn treated with DuPont™ Aproach® Prima fungicide at the VT, R1, or R2 growth stage to non-treated corn.

Southern Rust

Scientists conducted fungicide research trials over two years in the southeastern U.S. at locations where southern rust was the predominant foliar disease.

A replicated research study was conducted near Camilla, GA in 2014 to assess southern rust infestation and corn yield response of six different Pioneer® hybrids with and without foliar fungicide treatment. This study included two different fungicide treatments: a single application at the V8-V10 growth stage, as well as a two-pass program with applications at both the V8-V10 stage and the VT-R1 stage. (The original protocol called for only the VT-R1 application, but treatment timings were altered when southern rust was detected earlier than expected at the research site.) Averaged over hybrids, there was no yield increase with early fungicide application alone, but yields were increased by an average of 20 bu/acre when the VT-R1 application was included.

COTTON

Introduction

Cotton, one of the world's leading agricultural crops, is plentiful and economically produced, making cotton products relatively inexpensive. The fibres can be made into a wide variety of fabrics ranging from lightweight voiles and laces to heavy sailcloths and thick-piled velveteens, suitable for a great variety of wearing apparel, home furnishings, and industrial uses. Cotton fabrics can be extremely durable and resistant to abrasion.

Cotton accepts many dyes, is usually washable, and can be ironed at relatively high temperatures. It is comfortable to wear because it absorbs and releases moisture quickly. When warmth is desired, it can be napped, a process giving the fabric a downy surface. Various finishing processes have been developed to make cotton resistant to stains, water, and mildew; to increase resistance to wrinkling, thus reducing or eliminating the need for ironing; and to reduce shrinkage in laundering to not more than one percent.

Nonwoven cotton, made by fusing or bonding the fibres together, is useful for making disposable products to be used as towels, polishing cloths, tea bags, tablecloths, bandages, and disposable uniforms and sheets for hospital and other medical uses.

Cultivation of the cotton plant

The various species of cotton grown as agricultural crops are native to most subtropical parts of the world and were domesticated independently multiple times. Cotton can be found as perennial treelike plants in tropical climates but is normally cultivated as a shrubby annual in temperate climates. Whereas it grows up to 6 metres (20 feet) high in the tropics, it characteristically ranges from one to two metres (3 to 6.5 feet) in height under cultivation. Within 80–100 days after planting, the plant develops white blossoms, which change to a reddish colour. The fertilized blossoms fall off after a few days and are replaced by small green triangular pods, called bolls, that mature after a period of 55–80 days. During this period the seeds and their attached hairs develop within the boll, which increases considerably in size.

The seed hair, or cotton fibre, reaching a maximum length of about 6 cm (2.5 inches) in long-fibre varieties, is known as lint. Linters, fibres considerably shorter than the seed hair and more closely connected to the seed, come from a second growth beginning about 10 days after the first seed hairs begin to develop. When ripe, the boll bursts into a white, fluffy ball containing three to five cells, each having 7 to 10 seeds embedded in a mass of seed fibres.

Two-thirds of the weight of the seed cotton (i.e., the seed with the adhering seed hair) consists of the seeds. The fibres are composed of about 87 to 90 percent cellulose (a carbohydrate plant substance), 5 to 8 percent water, and 4 to 6 percent natural impurities.

Although cotton can be grown between latitudes 30° N and 30° S, yield and fibre quality are considerably influenced by climatic conditions, and best qualities are obtained with high moisture levels resulting from rainfall or irrigation during the growing season and a dry, warm season during the picking period.

To avoid damage to the cotton by wind or rain, it is picked as soon as the bolls open, but since the bolls do not all reach maturity simultaneously, an optimum time is chosen for harvesting by mechanical means. Handpicking, carried out over a period of several days, allows selection of the mature and opened bolls, so that a higher yield is possible.

Handpicking also produces considerably cleaner cotton; mechanical harvesters pick the bolls by suction, accumulating loose material, dust, and dirt, and cannot distinguish between good and discoloured cotton. A chemical defoliant is usually applied before mechanical picking to cause the plants to shed their leaves, thus encouraging more uniform ripening of the bolls.

Typically, the disease is initiated on leaves in the lower canopy. The disease then progresses up the plant resulting in defoliation of leaves that would normally contribute to boll development. Subtending leaves, the leaf closest to a square or boll on a fruiting branch, are a major source of photosynthates required for boll development.

If these subtending leaves are impacted by disease, it can affect boll development. The critical phase to protect against premature defoliation is during the flowering and boll development (Fig. 3). This disease is common, but does not cause significant yield loss every season.

Fig. 3. Characteristic lesions of target spot of cotton.

Growers most likely to see benefit from applying fungicide for target spot have a crop with 1) high yield potential, 2) excellent growth, 3) disease appears at or soon after first bloom, and 4) grower makes fungicide application by 3rd week of bloom.

Growers least likely to see a benefit from a fungicide application are those who a) have poor growth and a crop suffering from nutrient deficiency, b) apply fungicide where target spot never develops, c) apply fungicide after target spot has caused 25% or more defoliation.

From field research, we know that well timed fungicide applications can prevent significant yield loss. The general recommendation is to scout for disease and if you find target spot lesions developing before the sixth week of bloom, it is advisable to make a fungicide application and possibly a second application 14 days later, if conditions are favorable.

Fungicide applications at the first and third week of bloom have been shown to be optimal timing for prevention of premature defoliation. After the sixth week of bloom the economic return on fungicide application begins to decline.

Areolate Mildew

Areolate mildew is another foliar pathogen of cotton caused by *Ramulariopsis* species. It produces distinctive fungal growth similar to other mildews encountered on crops such as cucurbit mildews. This disease can be easily identified by the presence of white to grayish colored spores on the underside of the leaf (Fig. 4).

For some varieties the upper leaf surface of infected leaves turns a purple color. However, there is a notable distinction between areolate mildew and other mildews that we commonly think of, such as powdery mildew of pumpkin. Powdery mildew of pumpkin is known as an obligate parasite, meaning that it requires living host tissue to complete its life cycle.

If you were to try to grow the pathogen that causes powdery mildew on nutrient media in a petri dish it would not survive. Areolate mildew is also known as a false mildew, and it is able to grow on nutrient media without the presence of a living host.

Fig. 4. Areolate mildew visible on the underside of a cotton leaf.

Because of this ability to live on non-living residue, areolate mildew can persist on decomposing plant material. Areolate mildew can cause premature defoliation, but fortunately the disease can be managed with timely fungicide application.

Diagnosis and Management of Foliar Diseases

Name: Ascochyta Blight (Wet Weather Blight)

Caused by: Ascochyta gossypii (*Phoma exigua*)

Foliar Symptoms: Ascochyta Blight forms lesions on cotyledons, leaves, stems, and bolls. Lesions on the cotyledons and leaves approach 2 mm.

Management: Use fungicide treated seed, avoid planting in cool wet weather, and incorporate cotton residue to encourage decomposition, which helps in fields with a history of Ascochyta Blight.

Diagnostic Note: Margins of necrotic regions on leaves and cotyledons will have dark borders. Spots may have a target-like appearance. However, Ascochyta Blight typically occurs early in the season, and small black fruiting structures are observed in the lesions.

Range and Yield Loss: Ascochyta Blight has been reported in most major cotton producing regions. Yield loss is rarely reported, but is possible under conditions such as prolonged cloudy weather with cooler temperatures and rainfall.

Fig. 5 Symptoms of Asochyta Blight on cotton leaf.

Name: Bacterial Blight (Angular Leaf Spot, Black Arm)

Caused by: *Xanthomonas citri pv. malvacearum*

Foliar Symptoms: Symptoms of Bacte-rial Blight start as tiny water-soaked spots, and progress into characteristi-cally angular shapes due to leaf veins limiting bacterial movement. Lesions appear on the upper side of the leaf, turn black as they expand, and defoli-ation may occur. Systemic infections follow the main veins as black streaks; symptoms on the bolls are characteristically sunken, water-soaked lesions.

Management: Resistant cultivars are the most economical option to minimize yield loss. Incorporation of infected residue into the soil will help with decom-position of infected debris and reduce inoculum in the soil.

Diagnostic Note: Lesions are typically dark brown (darker than many other pathogens) and can be "shot-holed" in appearance when necrotic tissue falls away. Observe the underside of the leaf for water-soaking around the lesions.

Fig. 6 Dark-brown angular lesions and 'black arm' of peti-oles of Bacterial Blight.

Range and Yield Loss: Bacterial Blight is a major disease of cotton. Since acid delinting of cottonseed in the U.S., Bacterial Blight has been rare, except in OK and TX; however, there is a recent resurgence in additional states. Yield loss can be severe, up to 20%, depending on variety and pathogen race.

Name: Target Spot

Caused by: *Corynespora cassiicola*

Foliar Symptoms: Characteristic symptoms of Target Spot include brown lesions, sometimes approaching 2 cm (~1 inch) in diameter, exhib-iting a series of concentric rings. Unlike Stemphylium and Alternaria Leaf Spot, the spots are typically not bordered by a dark band. Leaf spots and premature defoliation are generally confined to the interior canopy (unlike that found in Stemphylium and Alternaria diseases.)

Management: Management of Target Spot with fungicides is made difficult by the dense canopy which obstructs placement of fungicides in the interior of the canopy. Some success in managing Target Spot has been achieved by applying fungicides during both the 1st and 3rd weeks of bloom.

Diagnostic Note: Target Spot can be quickly differentiated from Stemphylium Leaf Spot by location in the canopy. Also, defoliation from Target Spot typically begins at the bottom of the plant and progresses upwards. Lesions may occur in the upper canopy, but defoliation of the upper 20% has not been observed

Range and Yield Loss: Target Spot has become progressively more wide-spread in the Southeast and Mid-South regions of the Cotton Belt, but is most severe in Florida, Alabama and Georgia. In severe cases, yield losses exceeding 200 lbs lint/A have been documented

Name: Cercospora Leaf Spot

Caused by: *Cercospora gossypina, Mycosphaerella gossypina*

Fig. 7 Typical symptomatic lesion of Target Spot.

Foliar Symptoms: Reddish lesions will occur during the early stages. As the disease progresses, the lesions enlarge and turn white to light brown in the center. The lesions are circular and vary in size. Concentric zones are often present with a red color at the margins.

Management: Maintaining plant vigor by having proper fertility and preventing drought stress through irrigation helps delay primary infections and reduce the severity of disease outbreaks.

Diagnostic Note: In the field it is often difficult to differentiate Cercospora Leaf Spot from other foliar diseases.

The spots appear concentric, like those of Target Spot, and the general distribution of spots on the plant may be the same as Stemphylium and Alternaria Leaf Spots. Correct diagnosis often requires viewing the long, thin whip-like, septate spores. Caution should be taken as C. cassiicola spores can appear similar to Target Spot, but typically are broader and may have a basal scar.

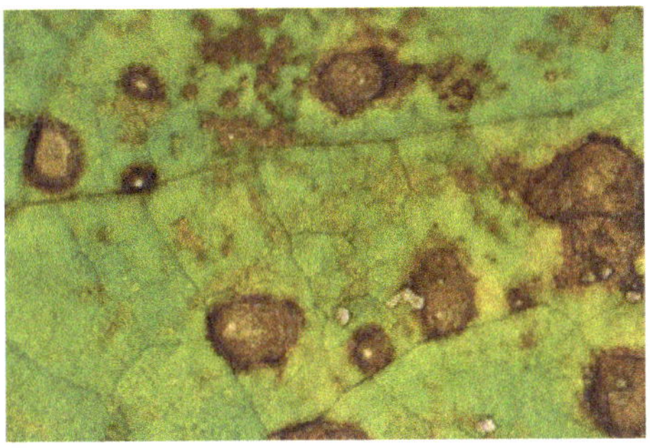

Fig. 8 Foliar symptoms of Cercospora Leaf Spot

Range and Yield Loss: Cercospora Leaf Spot occurs in all cotton producing areas within the U.S. When Cercospora Leaf Spot occurs while plants are under stress, or in a disease complex with Alternaria or Stemphylim Leaf Spot, premature defoliation, reduced yield, and lower fiber quality have been documented.

Name: Alternaria Leaf Spot

Caused by: *A. macrospora, A. alternata*

Foliar Symptoms: Alternaria Leaf Spot forms lesions on senescing leaves that are brown with purple margins. As lesions expand they typically exhibit concentric zonation and the necrotic tissue will overlap with other lesions. As the disease progresses the lesions will become gray and dry with some of the necrotic tissue falling out giving it a "shot-holed" appearance.

Management: Reducing plant stress and insuring proper soil fertility, especially with potassium, can reduce disease severity.

Diagnostic Note: Symptoms of Alternaria and Stemphylium Leaf Spots are similar; however, Alternaria Leaf Spot is more commonly observed in Texas and the Mid-South and Stemphylium Leaf Spot in the Southeast. Lesions with concentric rings may appear similar to Target Spot; however, spots from Alernaria Leaf Spot will occur throughout the canopy and are also associated with reddening and yellowing leaves.

Range and Yield Loss: Alternaria is one of the most common cotton diseases and is associated with late season cotton development. (Note: Alternaria and Stemphylium Leaf Spots are similar in that both are most severe when cotton plants have insufficient potassium, either from inadequate fertilization or dur-ing periods of drought.) Yield loss is not considered a problem with Alternaria Leaf Spot if symptoms and defoliation occur late in the season.

Fig. 9 Foliar symptoms of Alternaria leaf spot.

Name: Stemphylium Leaf Spot

Caused by: Stemphylium solani

Foliar Symptoms: Stemphylium Leaf Spot lesions are 2 cm in diameter (~1 inch), circular in shape and brown in color with concentric zones. As lesions mature they will develop a whitish center that may crack and fall out producing a "shot-holed" appearance. The lesions normally form on the up-per leaves in the canopy and start at the leaf margin and move inward. Fields where Stemphylium Leaf Spot is severe also typically demonstrate symptoms of nutrient deficiency.

Management: Managing for vigorous crop growth by irrigating to avoid drought stress, proper fertilization, and reducing pest pressure will help reduce the incidence of disease outbreaks. Fungicides are available, but are not normally economical for the control of Stempyhlium Leaf Spot.

Diagnostic Note: Stemphylium Leaf Spot can be differentiated from Target Spot by position on the plant (Stemphylium typically is found at the top of the plant first, Target Spot is within the canopy). Also, leaves affected by Stemphylium Leaf Spot often show nutrient deficiencies (yellow and red) while leaves affected by Target Spot often appear green, unless approaching senescence.

Range and Yield Loss: Stemphylium Leaf Spot has been found in all the cotton producing areas of the U.S. Since this disease is normally associated with other plant stress factors (drought, nutrient deficiencies, nematode and insect pressure) yield loss can be severe from the stress complex.

Name: Areolate Mildew

Caused by: Ramularia gossypii

Foliar Symptoms: Small lesions appear on leaves in the lower canopy late in the growing season. The lesions, 3-4 mm (~0.15 in) wide and restricted by a major leaf vein, are slightly chlorotic on the upper leaf surface with a white mildew growth on the lower surface. Lesions may become necrotic and resemble bacterial blight. In severe cases, premature defoliation will occur.

Fig. 10. Foliar symptoms of Stemphylium leaf spot.

Management: Use of resistant varieties is the best control strategy for Areolate Mildew. Incorporating crop residue and crop rotation will help reduce inoculum for next season. Fungicides (e.g. azoxystrobin and pyraclostrobin) were effective in managing this disease in Georgia.

Diagnostic Note: No other cotton disease commonly observed will have the white, powdery growth on the underside of the leaf.

Range and Yield Loss: Areolate Mildew is observed in most countries where cotton is produced, but not commonly observed in the U.S. In Georgia, the disease is typically restricted to the extreme southeastern counties. This disease normally appears late in the season and usually causes little yield loss.

GRAIN CROPS

Along with cultural practices, the main control measures for the management of fungal diseases are the application of effective fungicides and the use of crop varieties with genetic resistance. However, due to the lack of highly resistant varieties in the grains industry, fungicides are frequently solely relied upon for the control of many fungal diseases. In Australia, the fungicides used against grain diseases are predominantly of the azole or demethylase inhibitor (DMI) group 3, although there are other groups like the quinone outside inhibitor (QoI) group 11 and the succinate dehydrogenase inhibitor(SDHI) group

7 that also contribute to the control as existing mixing partners or new solo and mixed formulations. The threat of so few fungicide modes of action, is that when a fungal disease develops resistance to one fungicide all other fungicides which share the same mode of action, or within the same fungicide group, are also at risk.

Fig. 11. Symptoms of Areolate Mildew.

So far seven cases of fungicide resistance and two cases of fungicide tolerance in grain diseases have been detected in the last five years in Australia (Table 1); many horticultural crops have not been studied yet but anecdotal evidence suggests that many more cases remain to be uncovered.

Table 1. Fungicide resistance cases identified

Disease	Fungicide
Barley powdery mildew	DMI
Wheat powdery mildew	DMIc, strobilurins
Canola blacklega,	MAP/Kinase, DMI
Legume Botrytis	MBC
Lentil Ascochyt	MBC
Barley net-blotches	DMI
Wheat septoria leaf blotch	DMI
Grape powdery mildew	QoI/DMI
Grape downy mildew	PAA
Grape Botrytisa	MBC/E1/E3/D1

The azoles remain the backbone of crop protection in broad-acre cropping worldwide and no catastrophic cases of resistance have been reported anywhere in the world. For these reasons, the monitoring of fungicide resistance in Australian agriculture generally was considered unimportant until comparatively recently. Today we can see that the situation in Australia is at a cross-road, with significant levels of resistance in all studied crops.

Resistance in barley powdery mildew starts to develop in the east

DMI resistance in barley powdery mildew was first documented in 2009 in WA. The high levels of resistance detected against certain DMIs were associated with a number of mutations in the target site of this group of fungicides, the most important being mutations Y136F and S509T. Since 2009, great efforts have been made to understand the factors responsible for the rapid onset of this resistance and the spread of the problem (for details see previous paper presented at Bendigo updates titled "*From the lab to the field: The scale and impact of fungicide resistance in Australia*"), especially to other states across the country.

In 2012, the first of these two mutations, Y136F, was found in barley powdery mildew samples collected from Victoria. This mutation alone has only marginal effects on the sensitivity levels to the fungicides in laboratory conditions, but it is required for the development of further mutations which have been reported to show field resistance. This mutation is therefore what we call a gateway mutation. In 2015, and thanks to the implementation of a new detection technology used in cancer research, the second mutation, S509T, was subsequently detected in NSW, Victoria and Tasmania (Fig. 12).

In order to limit the potential impact of these mutations spreading across the barley growing areas in the Eastern states, we need to answer the following questions i) is the resistance coming from the West?, and ii) how are the resistant populations going to evolve considering the differences between barley crops in the East and the West in terms of pathogen selection pressure?

The answer to the first question is part of an ongoing research in which the analysis of DNA from both populations tries to determine if there is one centre of origin (West) or if the Western and Eastern fungicide resistant populations evolved and established independently. The second question is probably most important from the disease management point of view. Considering the chemical and cultivar alternatives available, and the existing cultural differences between both areas, it is difficult to envisage a situation similar to the 2009-2012 barley powdery mildew fungicide resistance outbreak recorded in WA. However, in order to reduce the spread of these resistant populations it is necessary to monitor and implement anti-resistance management strategies based on the use of i) chemistries from different modes of actions, ii) varieties with higher levels of resistance, and iii) the introduction of break crops to decrease the disease load from season to season.

Net type net blotch joins the fungicide resistance race

In Australia, emergence of multi-DMI resistance in net type net blotch has been observed in samples collected from the years 2013 onwards. Resistant isolates were geographically widely dispersed across Western Australia, originating in Kojonup, Beverley, Bakers Hill, West Arthur and Dandaragan. We used trials in these studies so as to increase the likelihood of finding resistant isolates even when their frequency is still low. We therefore cannot reliably estimate the frequency of the resistant net type net blotch isolates, but results so far indicate it is at a significant level in Western Australian populations.

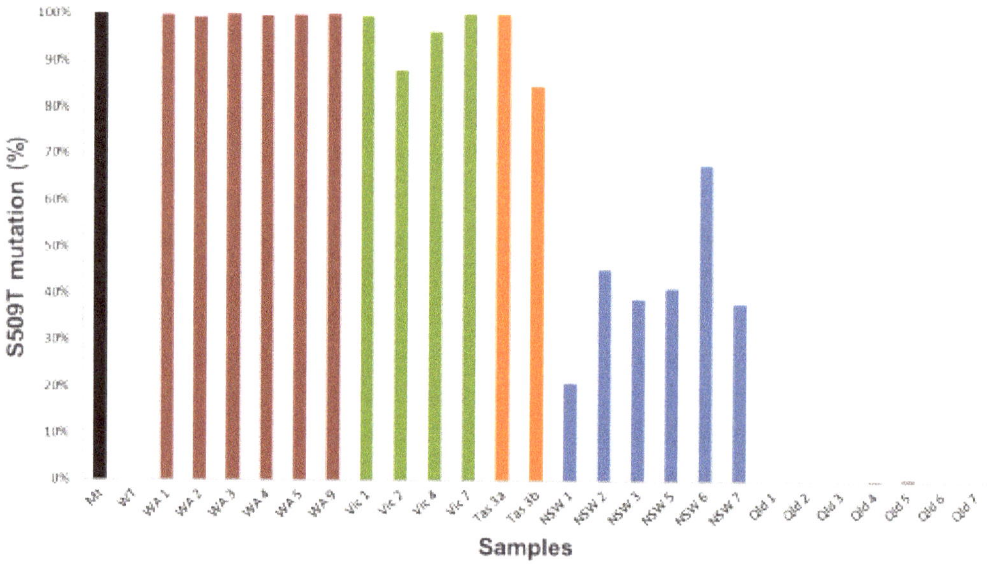

Fig. 12. Mutation rates in barley powdery mildew samples quantified by digital PCR in 2016 in Western Australia, Victoria, Tasmania, New South Wales and Queensland. Samples were obtained from barley bait trails specifically developed to catch resistant strains.

When the resistance mechanism was analysed, we found that a combination of mutations and increased expression of the fungicide target were responsible for this resistance in the fungus.

These changes were effectively increasing *in vitro* the resistance factor (how many times a resistant isolate is more resistant to a given fungicide compared to the average sensitive pathogen population) of the resistant strains when exposed to the fungicides epoxiconazole (RF=1.5), prothioconazole (RF=2.6) and propiconazole (RF=7.7). However, these changes were also increasing the resistance factors for non-registered fungicides such as difenoconazole (RF=12.4), tebuconazole (RF=16.5) and prochloraz (RF=27.7), which critically highlights the fact that fungicides belonging to the same mode of action group (DMIs) are often affected by this type of resistance due to the existing similarities in their structures even if they have not been exposed to the disease previously. So far no resistance has been found in any of the analysed samples from the Eastern States. However, it is important to mention that only a small number of net type net blotch samples were tested and that a large scale analysis is required to determine whether resistance has developed outside of Western Australia.

An important question that remains is the reason why resistance has not developed in spot type net blotch. Net and spot type are two diseases caused by two pathogens that, in addition to the host, share many genetic similarities and undergo similar selection pressures for resistance in the field. We believe that understanding the mechanisms responsible for resistance in net type can contribute to the development of better anti-resistance management practices that can help the industry minimise the risk of fungicide resistance in spot type.

Defining Fungicide Resistance

A potential new fungicide is identified in laboratory and glasshouse tests on different types of fungal pathogen, and is then tested in field trials against an appropriate range of crop diseases in different regions and countries. Only if it works uniformly well against important crop diseases in a large number of trials over several seasons is it considered for development and marketing. The pathogens it works against are deemed to be 'sensitive', and those that it does not affect or hardly affects are regarded as 'naturally' or 'inherently resistant'. This pre-existing type of resistance is of no further practical interest once it has been identified as a limitation to the range of use of the fungicide. Reasons for natural resistance are seldom investigated, although sometimes they can be deduced from mode of action studies.

The 'fungicide resistance' we are considering here is a different phenomenon, sometimes called 'acquired resistance'. Sooner or later during the years of commercial use of a fungicide, populations of the target pathogen can arise that are no longer sufficiently sensitive to be controlled adequately. They generally appear as a response to repeated use of the fungicide, or to repeated use of another fungicide which is related to it chemically and/or biochemically through a common mechanism of antifungal action. This emergence of resistant populations of target organisms, which were formerly well controlled, has been widely known for antibacterial drugs (e.g. sulphonamides, penicillin, streptomycin) and for agricultural and public health insecticides (e.g. DDT) for almost sixty years.

Some people prefer to call this phenomenon 'insensitivity' or 'tolerance'. The former term is preferred by some plant pathologists, because they believe that fungicide resistance is easily confused with host-plant resistance to certain species or pathotypes of fungi. Some agrochemical companies have also tended to use 'insensitivity', 'loss of sensitivity' or 'tolerance', because these sound less alarming than 'resistance'. On the other hand, two studies on terminology recommended that 'resistance' should be the preferred term. Also 'resistance' has been in use for many years to describe precisely the same phenomenon in bacteriology and entomology, and it is now very widely used with reference to fungicides also.

Workers within the agrochemical industry have objected from time to time to the use of 'resistance' to describe shifts in fungicide sensitivity occurring either in non-crop situations such as the laboratory or experimental glasshouse, or in the field but to a degree which is too small to affect disease control. They recommend that 'resistance' should denote only situations where failure or diminution of crop disease control is known to have resulted from a change in sensitivity. It is true that observations of 'resistance' generated in the laboratory, and detection of rare or weakly resistant variants in the field, have on occasions been misinterpreted by scientific authors, or by commercial competitors, as indicating actual or impending failure of a product to perform in practice, when in fact good control was still secured.

However, attempts to restrict in this way the meaning of such a broadly used term as 'resistance' are bound to fail and to create more confusion. It is better to qualify the term when necessary. 'Field resistance' (in contrast to 'laboratory resistance') has been used sometimes to denote specifically a crop disease control problem caused by resistance. However, detection of some signs of resistance in the field can still be a far cry from having a control failure. It seems preferable to use 'field resistance' to indicate merely the presence of resistant variants in field populations (at whatever frequency or severity), and 'practical resistance' to indicate consequent, observable loss of disease control, whenever such precise terminology is necessary. 'Laboratory resistance' or 'artificially induced resistance' also are useful, precise terms which are self-explanatory. Some authors have claimed to find 'field resistance' in studies where the resistant variants actually were detected only after the field samples were subjected to subsequent selection by exposure to the fungicide in the laboratory. This is a borderline case, which is hard to categorise.

Occurrence of resistance

Leading examples are given of the more important diseases affected, and a few key literature references are cited. Up to 1970 there were a few sporadic cases of fungicide resistance, which had occurred many years after the fungicide concerned was introduced. With the introduction of the systemic fungicides, the incidence of resistance increased greatly, and the time taken for resistance to emerge was often relatively short, sometimes within two years of first commercial introduction. Many of the fungicides introduced since the late 1960s have been seriously affected, with the notable exceptions of the amine fungicides ('morpholines'), fosetyl-aluminium, anilinopyrimidines, phenylpyrroles and some of the fungicides used to control rice blast disease (e.g. probenazole, isoprothiolane and tricyclazole), which have retained effectiveness over many years of widespread use. Some recently introduced fungicides such as benzamides and carboxylic acid amides have not yet encountered serious resistance problems, possibly because of the management precautions which have been taken. Most of the older materials such as copper fungicides, sulphur, dithiocarbamates (e.g. mancozeb), phthalimides (e.g. captan) and chlorothalonil, have retained their full effectiveness in all their uses, despite their extensive and sometimes exclusive use over many years.

Often the onset of resistance has been associated with total, or almost total, failure of disease control. Indeed it was growers' observations of obvious and sudden loss of effect that generally gave the first indication of resistance. Of course it was necessary to show that resistance really was the cause, by checking for abnormally low sensitivity of the pathogen in tests under controlled conditions. There was, and to some extent still is, a temptation for growers and advisers to blame resistance for all cases of difficulty of disease control. There are many other possible reasons, such as poor application, deteriorated product, misidentification of the pathogen, unusually heavy disease pressure. However, there remained many examples where no other explanation was found, and where serious loss of control was clearly correlated with greatly decreased sensitivity of the pathogen population as revealed in laboratory tests on representative samples.

Resistance of the kind just described, characterised by a sudden and marked loss of effectiveness, and by the presence of clearcut sensitive and resistant pathogen populations with widely differing responses, is variously referred to as 'qualitative', 'single-step', 'discrete', 'disruptive' or 'discontinuous' resistance. Once developed, it tends to be stable. If the fungicide concerned is withdrawn or used much less, pathogen populations can remain resistant for many years; a well-documented example is the sustained resistance of Cercospora betae, the cause of sugar-beet leafspot, to benzimidazole fungicides in Greece. A gradual recovery of sensitivity can sometimes occur, as in the resistance of Phytophthora infestans, the potato late blight pathogen, to phenylamide fungicides. In such cases, resistance tends to return quickly if unrestricted use of the fungicide is resumed, but re-entry involving also a partner fungicide has proved useful in some instances.

Sometimes, as in the case of the DMI fungicides, and of the 2-amino-pyrimidine fungicide ethirimol, resistance has developed less suddenly. In such cases, both a decline in disease control and a decrease in sensitivity of pathogen populations as revealed by monitoring tests, manifest themselves gradually, and are partial and variable in degree. This type of resistance is referred to as 'quantitative', 'multi-step', 'continuous', 'directional' or 'progressive'. It reverts rapidly to a more sensitive condition under circumstances where the fungicide concerned becomes less intensively used and alternative fungicides are applied against the same disease. The first appearance of resistance in a particular fungicide-pathogen combination in one region has almost always been accompanied, or soon followed, by parallel behaviour in other regions where the fungicide is applied at a similar intensity. Whether the fungicide also meets resistance in other of its target pathogens depends on the individual case. Generally it does occur in other target pathogens that have a comparable rate of multiplication, provided that the fungicide is used in an equally intensive way. It is notable that rust fungi, despite their abundant sporulation and rapid spread, appear to be low-risk, seldom producing resistance problems.

Pathogen populations that develop resistance to one fungicide automatically and simultaneously become resistant to those other fungicides that are affected by the same gene mutation and the same resistance mechanism. Generally these have proved to be fungicides that bear an obvious chemical relationship to the first fungicide, or which have a similar mechanism of fungitoxicity. This is the phenomenon known as 'cross-resistance'. For example, pathogen strains that resist benomyl are almost always highly resistant to other benzimidazole fungicides such as carbendazim, thiophanate-methyl or thiabendazole. Sometimes cross-resistance is partial, even when allowance is made for the greater inherent activity of different members of a fungicide group.

There is a converse phenomenon, 'negative cross-resistance', in which a change to resistance to one fungicide automatically confers a change to sensitivity to another. This is much rarer, but several cases are well characterised; one, involving carbendazim and diethofencarb, has been of practical importance and is discussed later.

Some pathogen strains are found to have developed separate mechanisms of resistance to two or more unrelated fungicides. These arise from independent mutations that are selected by exposure to each of the fungicides concerned. This phenomenon is totally different from cross-resistance in its origin and mechanism, and is usually termed 'multiple resistance'. An example is the common occurrence of strains of Botrytis cinerea that have become resistant to both benzimidazole and dicarboximide fungicides.

Origins of resistance

Once it arises, resistance is heritable. It results from one or more changes in the genetic constitution of the pathogen population. There is overwhelming circumstantial evidence that a mutant gene that causes production of a particular resistance mechanism pre-exists in minute amounts in the population. Before the fungicide was ever used in the field, such a mutation would confer no advantage to the growth or survival of the organism, and could well cause a slight disadvantage. Hence it would remain at a very low frequency, probably dying out and re-appearing spontaneously many times.

Spontaneous mutations of all kinds are continually occurring in all living organisms. The rate of mutation can be increased greatly in the laboratory by exposing the organism to ultra-violet light or chemical mutagenic agents, and thus resistant mutants can be produced artificially. However, it cannot be assumed that such artificial mutants are necessarily identical in resistance mechanism or in other respects to those that arise in the field.

Typically, a resistant mutant might exist at an initial frequency of the order of 1 in 1000 million spores or other propagules of the pathogen. Amongst the survivors of a fungicide treatment, however, the resistant forms will be in much higher proportion ('the survival of the fittest'). It is only when this reaches say 1 in 100 or even 1 in 10 in the population that difficulty of disease control and the presence of resistant individuals will have become readily detectable.

Thus the obvious onset of resistance is often sudden, but prior to this the resistance will have been building up insidiously at undetectable levels. If a fungicide treatment is very effective, with few survivors, selection will be very rapid. If the fungicide is only

80% effective, then after each treatment the number of variants will be concentrated only 5-fold and the build-up will be slower.

Several fairly obvious but important deductions, which can influence assessment of risk and design of avoidance strategies, can be made from consideration of this simple process of mutation and selection. Accumulation of resistant mutants will be enhanced by higher frequency of treatment with the fungicide concerned, by a more effective application method or dose, by the presence of larger pathogen populations before treatment, and by greater spore production and shorter generation times in the pathogen.

The selection process outlined above is based on much genetic analysis of sensitive and resistant strains, and on much field experience. However, it represents the simplest form of resistance, the discrete pattern referred to earlier, which is also termed 'major gene' resistance.

One point mutation causing a single amino acid change in the target protein is responsible for a high level of resistance, and the sensitive and resistant forms fall into very distinct classes. This pattern is characteristic of resistance to several major groups of fungicides including benzimidazoles, phenylamides, dicarboximides and QoIs.

Other mutations in the target protein may give rise to lower levels of resistance. For example, the F129L mutation in the b-cytochrome target of QoIs causes only low levels of resistance in many pathogens, and hence is of little practical importance, in contrast to the G143A mutation which causes a high degree of resistance, and consequent loss of disease control.

A somewhat different 'polygenic' process of genetic change is thought to underlie the 'quantitative' or 'multi-step' pattern of resistance. Again resistance results from the selection of mutants, but in this case a number of different genes, each with a partial effect, appear to be involved.

The more genes that mutate to resistance-causing forms, the greater the degree of resistance. This would account for the gradual observable development of resistance, and for the continuous range of sensitivity that can be found (Fig.12).

Although the theory of polygenic resistance is widely accepted, it must be said that the genetic evidence for polygenic resistance in field isolates is rather thin. The best known and most studied examples of continuous resistance in practice have been in cereal powdery mildews, which are rather hard to study genetically, and some of the data are conflicting. Biochemical evidence for polygenic resistance to azole (DMI) fungicides indicates involvement of at least four resistance mechanisms which are discussed below.

However, researchers studying the human pathogen Candida albicans, found that different mutations in the same target-site gene may accumulate in a single strain, and their individual effects may be additive, or possibly synergistic. In this way polyallelic changes may contribute to multistep development of resistance.

QoIs (strobilurins) are the first fungicide class to target a protein (cytochrome bc-1) that is encoded by a mitochondrial gene. DNA repair mechanisms are less effective for mitochondrial DNA than for nuclear DNA, and consequently mitochondrially encoded genes are more liable to mutation. The frequency of DNA base changes in mitochondrial DNA is further increased by its close proximity to reactive oxygen species generated during respiration. Depending on the impact of these mutations on fitness, resistance seems likely to develop quickly where target sites are encoded by mitochondrial genes. Onset of resistance to QoIs was in fact rapid in a number of pathogens, although it must be noted that benzimidazole resistance, resulting from a nuclear mutation, developed equally quickly.

Resistance Mechanisms

A large amount of experimental effort has focussed on this subject, particularly in academic laboratories. Some of the information is derived from resistant strains generated in the laboratory (e.g. for quinoxyfen) and not from field isolates. We now understand well the most important mechanisms of resistance to the benzimidazole, carboxanilide, phosphorothiolate, dicarboximide, and QoI fungicides.

There is extensive information concerning the DMI fungicides, identifying four major resistance mechanisms that may operate. However, there are still many gaps in our knowledge, not only for established fungicide groups (e.g. anilinopyrimidines), but also for new fungicide groups defined by cross-resistance (e.g. carboxylic acid amides, CAAs).

Many types of resistance mechanism are known. These include: alteration of the biochemical target site so that it is no longer sensitive; increased production of the target protein; developing an alternative metabolic pathway that bypasses the target site; metabolic breakdown of the fungicide; exclusion or expulsion of the fungicide through ATP-ase dependent transporter proteins.

By far the commonest mechanism appears to be an alteration to the biochemical target site of the fungicide. This could explain why many of the older products have not encountered resistance problems. Once they have penetrated the fungal cell, the older fungicides act as general enzyme inhibitors, affecting many target sites (hence they are sometimes called 'multi-site' inhibitors). They act selectively on fungi, rather than on plants and animals, because they penetrate and accumulate much more readily in fungi.

Many sites in the fungus would have to change simultaneously in order to stop the fungicide from working. The chances of the many necessary genetic changes happening are negligible, and in any case an organism with so many alterations would be highly unlikely to be pathogenic or even viable. The occasional cases of resistance to multi-site fungicides presumably have resulted from other types of mechanism, not involving the sites of action.

Table 2 Mechanisms of Fungicide Resistance

Fungicide or fungicide class	Mechanism of resistance
Aromatic	Unknown, but show cross-resistance with dicarboximides hydrocarbons and phenylpyrroles
Organo-mercurials	*Detoxification by binding substances
Dodine	Unknown
Benzimidazoles	Altered target site (ß-tubulin)
2-Amino-pyrimidines	Unknown
Kasugamycin	Altered target site (ribosomes)
Phosphorothiolates	Metabolic detoxification
Phenylamides	Possibly altered target site (RNA polymerase)
Dicarboximides and Phenylpyrroles in	*Altered target site (protein kinase involved
osmoregulation)	
DMIs	Increased efflux; altered target site; decreased demand for target-site product; target-site over-production
Carboxanilides	Altered target site (succinate-ubiquinone oxidoreductase)
QoIs (strobilurins)	Altered target site (ubiquinol-cytochrome c reductase)
Melanin Biosynthesis	Altered target site (scytalone dehydratase)
Inhibitors (Dehydratase)	MBI-D

**Some doubt regarding occurrence in field isolates.*

In contrast, modern fungicides act primarily at single target sites, and are often referred to as 'single-site' or 'site-specific' fungicides. Thus just a single gene mutation can cause the target site to alter, so as to become much less vulnerable to the fungicide. The rapid development over the past 10 years of PCR-based diagnostic methods for detection of point mutations causing resistance has aided the identification of resistance mechanisms, especially those involving target site changes.

Several major resistance genes have now been isolated and characterised. In each case a single point mutation causes a change in a single amino acid in the target protein so that the fungicide no longer binds so tightly. Different amino acid changes in a target protein can cause different levels of resistance.

For instance, as mentioned earlier, the G143A mutation (causing glycine to be replaced by alanine) at amino acid position 143 in the b-cytochrome of mitochondrial Complex III, causes higher levels of resistance to QoIs than the less common F129L mutation.

The way in which polygenic systems operate to give different degrees of resistance are less clearly understood. The relatively low level of resistance caused by each gene makes the mechanisms of resistance particularly hard to determine. In the case of the DMI fungicides there is some evidence that mutation of different genes may elicit a number of different resistance mechanisms listed in Table 2.

These are unrelated, but can act simultaneously and possibly in a synergistic way. It is interesting that those few fungicides that are not directly fungitoxic, but which act indirectly by affecting defence mechanisms in the host plant, e.g. probenazole, have not encountered resistance. Reasons for this are not clear.

POTATO

The potato was originated in the hills of Andes and Bolivia in South America, subsequently it was introduced into Europe by Spaniards in the second half of the 16th century, from there it spread throughout Europe and rest of the world in the mid 17th to mid of 18th century. In Asia, particularly in India, it was introduced by Portuguese in 17th century.

The late blight fungus co-evolved with potato in Central and South America and subsequently spread to other parts of the world mainly through infected seed tubers. The late blight disease caused by oomycete, which was initially reported as *Botrytis infestans* in 1845 by C. Montagne, later on German scientist Anton de Bary renamed as *Phytophthora infestans* (Mont.) de Bary.

The entire potato crop across Europe, especially in Ireland, was killed prematurely during 1844–45; leading to worst ever famine the 'Irish Potato Famine'. One million people died of starvation due to that famine and another million migrated to the USA and other parts of the world.

Potatoes are a staple crop worldwide with large scale production and over 5000 varieties in total. One of the reasons for the success and utilisation of potatoes as a staple crop is that they produce more food on less land faster than almost any other food crop with good climate adaptability and also containing significantly higher important proteins than grains like wheat.

Currently, potatoes stand as the fourth most important staple food crop for supplying global caloric intake behind wheat, rice and corn so optimal production is essential for a growing global population. Whilst recent media articles have noted dangers surrounding cooking potatoes at elevated temperatures, due to the production of carcinogenic acrylamide this is not a unique phenomenon to potatoes alone and cannot offset the positive dietary nutrients provided in the form of antioxidants, vitamins, polyphenols, minerals, carbohydrates, and dietary fibres. 2020 UK production figures for potatoes were estimated at 5.37 Megaton (million tons) despite lower areas planted than previous years, thus higher yields per area are being achieved.

Wales (combined with the West midlands) accounts for around 12.5% of the total planted area and total production of potatoes across the UK a significant proportion despite the country's relatively small size and high focus on livestock over arable.

The late blight disease was recorded in India for the first time between 1870 and 1880 in the Nilgiri hills. Under subtropical plains particularly in eastern part of the India, it was first observed in 1898–1900 in Hooghly district of West Bengal. In the northern part, it appeared for the first time in 1883 in Darjeeling and subsequently spread rapidly to other adjoining hills. The late blight disease was observed in Khasi hills (North-eastern Region) in 1885, Kumaon hills in 1897 and in Shimla hills (North-western Region) in 1902. During 1913, it appeared at several places in Assam and Bihar. In plains of Uttar Pradesh, it was reported for the first time in 1943 in Dehradun and Meerut.

Severe attack of the late blight was observed in Meerut district in 1949, 1950 and 1951 and subsequently in many other districts of Uttar Pradesh. In Punjab, the disease was occurred annually from 1958 to 1963 except during 1961. Potatoes had been grown in Mahabaleshwar hills and other parts of Maharashtra but late blight was observed there only in 1973. In Gujarat and Madhya Pradesh, the disease was observed in traces in 1968 and in Rajasthan in 1958. Afterwards, appearance of late blight disease is regular feature with high disease severity in hill areas while in plains disease severity is moderate to high level.

Disease impacts

Globally the FAO has suggested that 40% of yields across all crops are lost due to plant pests and diseases. Environmental changes with warmer temperatures which are set to occur due to the impacts of climate change are only likely to increase the risk of diseases and pests, as such this is a vital issue moving forwards in agriculture. This is particularly true of potato fungal pathogens as they require specific temperatures and humidity to thrive and predictions of climate change in the UK suggest this could cause increased blight issues. Across potato production estimates of 20% crop losses have been associated with late blight causing more than £3.5 billion losses worldwide. Late blight is considered the most important UK disease to potato production with estimated industry costs of £50 million per blight season. As such prevention and control of this disease is an important consideration in science and the sector as a whole. The disease is caused by the *Phytophthora infestans* a water mould fungus that can also cause pathogenic diseases in tomato crops (increasing its economic importance), with the main control being fungicide applications. This pathogen causes potato leaf and stem damage and eventual death which in turn causes the subterranean tuber to rot. *P. infestans* is particularly difficult to control as it undergoes rapid evolution via clonal reproduction and sexual reproduction and as such can develop resistances to treatments very quickly and particularly effective strains can dominate quickly. Many evolving strains of *P. infestans* are not only resistant to several fungicides and other biocides but they have also shown increased virulence and "aggression" causing increasingly worse damage to plants.

The late blight disease affects all plant parts especially leaves, stem and tubers

Late blight is caused by *Phytophthora infestans*, an oomycete. This is important to know, because fungicides that control true fungi do not always control oomycetes, and *vice versa*.

Leaves

Pale green water soaked spots (2–10 mm) appear mostly on the margin and tips. In moist weather, spots may appear anywhere on the leaves, enlarge rapidly and turn necrotic and black killing the entire leaf instantly. On the corresponding lower side, whitish cottony growth containing millions of sporangia forms around the dead area in a ring pattern (Fig. 14).

Symptoms on leaves: Round, watery-looking, dark brown blotches. Sometimes these blotches are surrounded by a yellowish-green ring.

Stem and petiole

Light brown lesions develop which elongates and encircles the stem and petioles breaking them and killing the plant/leaves instantly. Stem infection is more severe under high temperature and relative humidity conditions (Figure 15). Symptoms of stem blight are observed more in last ten years.

Tubers

Rusty brown discoloration of the flesh is the typical symptom of late blight (Fig. 16). On outside tuber surface, hard depressions with purplish tinge on the sides are a common feature. Normally, late blight infected tubers are hard but associated secondary pathogens may set in soft rot symptoms.

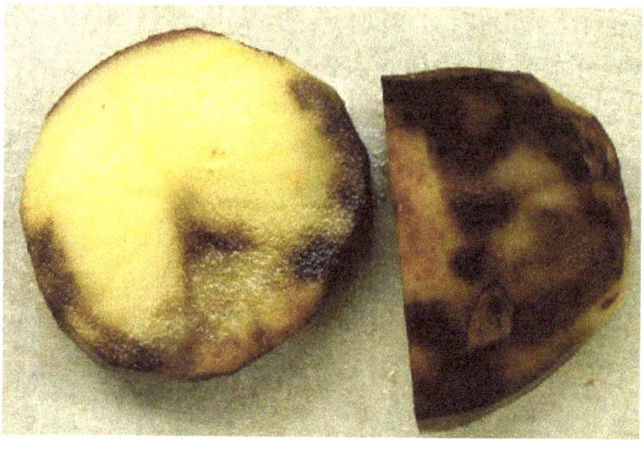

Fig. 13 Late blight (Phytophthora infestans)

Field infection

Generally, late blight appears on lower most leaves of the plant which goes unnoticed from a distance. Slowly, the disease spreads to the middle and then upper leaves. Subsequently it spreads whole plants and near of the plants. The disease spreads faster and the entire crop gets killed as if burnt by fire. The heavily infected field gives fetid odor which can be felt from a distance.

Fig. 14

Management

Several management strategies have been developed for late blight of potato and adopted by the farmers/potato growers as per availability of the resources. Amongst them chemicals, host resistant, biological control, cultural control are discussed below:

Fig. 15. Late blight symptom on potato stem.

Chemical management

Chemical management is very popular strategy for the management of late blight. Since the discovery of Bordeaux mixture in 1885 and it was first important landmark in the history of chemical disease control. Bordeaux mixture belongs to first generation of fungicides along with other inorganic chemicals. After more than 130 years, the introduction of Bordeaux mixture (Copper sulfate, hydrated lime and water), large numbers of fungicides (first generation Bordeaux mixture to fourth generation Mandipropamid & Azoxystrobin) were evaluated at worldwide against late blight of potato/tomato. In practice, the traditional management of late blight depends highly on preventative fungicides, application on a regular calendar basis (e.g. weekly) during the growing season.

Fig. 16. Late blight symptom on potato tubers.

The population diversity and disease incidence of *P. infestans* has been increased through the development of systemic fungicide resistance (insensitivity) and the transcontinental shipment of the late blight infected potato tubers and tomato plantlets. Metalaxyl fungicide which comes under Phenylamide group with FARC 4, was introduced against oomycetes, very effective for late blight management and highly adopted worldwide. However, after introduction within three years metalaxyl resistant isolates were detected on field grown potatoes. The site-specific systemic fungicide, mefenoxam (the active isomer in metalaxyl), inhibits sporulation and mycelial growth inside host tissues by specifically inhibiting RNA polymerase-1, a mutation that changes the affinity of target sites could easily lead to fungicide resistance.

In Indian scenario, metalaxyl based fungicides were introduced on experimental basis for management of late blight during late 1980's however, their commercial use started only during 1994–1995. In India, 200–400 ppm tolerance level was observed with metalaxyl. After 12 years, its introduction during 2006, the metalaxyl based fungicides failed to protect the potato crop from the late blight in temperate highlands leading to 40–70% crop losses. Systemic fungicide metalaxyl is cause of concern for management of late blight disease due to quickly developed resistance. Pathogen had developed.

Tolerance up to 400 ppm and genetic studies crosses indicated that a semi dominant major locus determines resistance to metalaxyl, since insensitive and sensitive parents usually yielded progeny with those phenotypes at a 1:1 ratio. The heterothallic single mating type isolates of *P. infestans* was exposed to 9 of the 11 commercial fungicide formulations for assess their effect on formation of oospores. The highest numbers of oospores were observed on media amended with Ridomil 2E (metalaxyl) and Ridomil Gold EC (mefenoxam) at 0.1 to 10 μg a.i./ml, when averaging it was found that 471 and 450 oospores/petri dish, respectively. The remaining fungicides viz., Maneb, Manzate (Mancozeb), Curzate (cymoxanil +mancozeb), and Acrobat MZ (dimethomorph + mancozeb) also induced oospore formation, which ranged from 0 to 200 oospores/petri at fungicide concentrations from 0.1 to 10 μg a.i./ml.

No oospores were formed on media amended with Bravo (chlorothalonil) or Tattoo C (chlorothalonil + propamocarb HCl), moreover both the compounds completely suppressed growth of the isolates at 0.1 and 1 µg a.i./ml. The metalaxyl resistant isolates formed oospores in response to the fungicides more often than the metalaxyl sensitive isolates. Metalaxyl + mancozeb (Ridomil MZ) and ofurace (Orafce 50WP) were reported to provide highly effective control of late blight. The fenamidone is a novel fungicide, which acts on cytochrome bc1 in mitochondrial complex III of *P. infestans* at a number of points in its life cycle. Cymoxanil based fungicides possess a novel mode of action by preventing electron transfer between cytochrome b and c1 in mitochondrial complex III and provide good scope for the control of late blight of potato and tomato.

Efficacy of seven fungicides was tested under *in vitro* conditions and the fungicides, which showed promising results, were further evaluated under field conditions and fenamidone based fungicide was found most effective in controlling late blight followed by cymoxanil based while mancozeb was found least effective; similarly the systemic fungicides viz., fenamidone and dimethomorph were reported most effective *in vitro* for management of late blight. Various studies showed that a reduced use of fungicides lowers the selection pressure for mefenaxam-resistant strains and mixture with a contact fungicide improves efficacy and may slow the development of resistance to mefenoxam.

The systemic fungicides have better persistence on the host surface and are being used as mixture with contact fungicides against late blight so as to avoid development of resistance in pathogen. The fungicide mixtures, containing two or more fungicides with different modes of action, have been developed with the twin objectives of broadening the activity spectrum against diverse plant diseases and to check the development of resistance in the target pathogens. In commercial production of potato is not viable without fungicides for management of late blight. Fungicide mixtures and targeted application based on late blight forecasting model are very important for managing late blight.

However, due to delisting of many fungicides products under the EU Pesticide Directive and environmental concerns, provides impetus for potato breeding and more effective fungicide application. It has been reported from European country that the same fungicide should not applied more than two sequential applications.

The severe late blight can be effectively managed with prophylactic spray of mancozeb at 0.25% followed by cymoxanil+mancozeb or dimethomorph + mancozeb at 0.3% at the onset of disease and one more spray of mancozeb at 0.25% seven days after application of systemic fungicides in West Bengal. Similarly, one spray of mancozeb followed by three spray of cymoxanil + mancozeb was effective on cv. Kufri Bahar under western UP.

Due to development of resistance to fungicides, a new fungicide, Victory 72 WP was first used in controlling late blight of potato and tomato in West Shoa of Ethiopia. The late blight specific spray scheduling method and a method of scheduling sprays for both diseases (early and late blight) suppressed early and late blight as well as did weekly sprays (conventional methods) and with the same average number of applications as with weekly sprays. The customarily, spray schedules were one prophylactic spray using contact fungicides followed by systemic fungicides and one more spray of either same contact or same systemic fungicides. A unique combination of treatments was developed keeping in view the sensitivity of *P. infestans* to develop fungicide resistance.

The post spray (curative spray) of same mode of action fungicide was not taken. Prophylactic sprays of chlorothalonil/mancozeb followed by systemic/trans laminar fungicides were found effective than post symptom sprays. This will be useful to minimize the yield losses due to late blight and assist in reducing development of resistance against fungicides in pathogen. The spray schedule of mancozeb 75% WP (0.2%- before appearance) followed by two more spray with mancozeb 75% WP (0.2%) + dimethomorph 50% WP (0.2%) at 7–10 days intervals showed less terminal disease severity (24.55%) with highest disease controlled (74.45%), which was at statistically par with treatment mancozeb 75% WP (0.2%, before appearance) followed by cymoxanil 8% + mancozeb 64% WP (0.3%) with two more spray at 7–10 days intervals, with 27.56% terminal disease severity along with disease controlled 71.29%.

One spray of mancozeb (contact fungicides: before appearance) and latter two more sprays of translaminar/systemic + contact fungicides at 7–10 days interval give better results for managing late blight of potato. The highest marginal benefit was achieved by applying first Ridomil then Dithane M-45 at 14–21 days interval. The lowest marginal benefit was with alone application of Ridomil at 21 day spray interval. At 7 days sprays was more economical to apply Dithane M-45 than Ridomil first followed by Dithane M-45 subsequently. Twelve fungicides were evaluated on isolates of three identified clonal lineages (US-22, US-23, and US-24) of *P. infestans* using a detached tomato leaf assay in preventative and post-infection methods.

The results revealed that these fungicides were suitable in conventional and organic systems, which can effectively control late blight caused by new clonal lineages of *P. infestans* when applied preventatively and late blight caused by the US-24 clonal lineage may require less fungicide than US-22 or US-23 to manage the disease. The efficacy of Ametoctradin 27% + dimethomorph 20.27% (w/w) as a new molecule for management of late blight of potato was reported in India.

Initium (ametoctradin) is a new fungicide for management of *Phytophthora infestans*. It affects mitochondrial respiration inhibitor interfering with the complex III (complex bc1) in the electron transport chain of the pathogen, thus ATP synthesis in the fungal cells is inhibited. It is a non-systemic fungicide that remains primarily on the leaf surface where it is adsorbed with high affinity to the epicuticular wax layer of the epidermis. Many oomycete-specific fungicides such as QoI compounds, dimethomorph, propamocarb, etc. were commercialized, but currently, we are unaware of any fungicide that could effectively halt epidemics caused by metalaxyl-resistant strains under conditions favorable to *P. infestans* growth and development.

Isolates of *Phytophthora infestans* showed 10-fold or more variation in baseline sensitivity to many fungicides including cymoxanil, dithiocarbamates, mandipropamid, and strobilurins. Various substances other than fungicides also were tested for management of late blight of potato. Ammonium molybdate, cupric sulfate and potassium metabisulfate at 1 mM partially inhibited the growth and spore germination of *P. infestans*, whereas ferric chloride, ferrous ammonium sulfate and $ZnSO_4$ at 10 mM completely inhibited growth and spore germination.

The foliar spray of $ZnSO_4$ and $CuSO_4$ (0.2%) micronutrients, 12 days delayed the onset of late blight when used with host resistance, subsequently reduced disease severity with higher yield. Sub-phytotoixc dose of boron with reduced rate of propineb + iprovalidicarb has been found more effective than treated with fungicides alone. □-aminobutyric acid (BABA) has been known as an inducer of disease-resistance. However, only the R but not the S enantiomer of BABA primes for resistance. Unfortunately, BABA can also impose growth stress in some treated plants therefore BABA analogs with reduced stress effects are highly desirable for agricultural field.

Plant activator viz., BABA and phosphoric acid was evaluated against late blight by various researchers with combination of fungicides or alone. A 20–25% reduction of the fungicide dose in combination with BABA gave on average the same result on late blight development as full dose Shirlan alone in field condition, while reduced dose of Shirlan alone sometimes resulted in less effective protection. However, *in vitro* results indicated that the efficacy was lasted for only 4–5 days after BABA treatment and subsequently efficacy was lowered. The partially resistant cultivars Ovatio and Superb reacted to lower concentrations of BABA where no effect was found in susceptible cv. Bintje.

Two SAR activators (BABA and phosphorous acid) were found effective against late blight of potato with significantly reduced disease severity (40–60%). The expression of the defense related genes and *P. infestans* effecter proteins □-1,3 glucanase, PR-1 protein, *phytophthora* inhibitor, protease inhibitor, xyloglucanase, thaumatin protein, steroid binding proteins, proline, endochitinase and cyclophilin genes were up regulated with the SAR activator treatment compared to unsprayed. Since last one and half decades, various fungicides have been developed for management of late blight. Isolates of *P. infestans* might develop resistant over the period. Fungicides resistance with currently used fungicides, including dimethomorph, has been reported. There are three key phases in the development of fungicide resistance (i) emergence, (ii) selection, and (iii) adjustment.

In emergence, the resistant strain has to arise through mutation and invasion whereas in selection, the resistant strain is present in the pathogen population and a small portion of the pathogen population carrying the resistance increases due to the selective pressure imposed by the fungicides. In case of adjustment phase, the resistant fraction of the pathogen population has become large, crop managers have to adjust fungicide programs, by changing the dose or active substance(s) used, in order to maintain control.

Biological control/eco-friendly management

Generally, management of late blight by eco-friendly means is a difficult task particularly when the level of disease pressure is high along with prevailing congenial environmental condition. However, due to negative impact of chemicals on environment as well as human health, nowaday's eco-friendly management is gaining more importance. Management of late blight through eco-friendly way, using botanicals has been initiated in European and American countries during the last years of 20th century.

Out of 100 species in 54 plant families tested against *P. infestans*, the leaf extracts from onions, garlic, *Malustoringo*, *Reynoutria japonica* and *Rheum coreanum* revealed positive inhibition of mycelial growth of *P. infestans*. *M. toringo* extracts strongly inhibited *P. infestans* and was effective in managing late blight also. The effectiveness of some antifungal compound was reported against late blight from botanicals.

The antagonist *Bacillus subtilis* B5 was found effective in inhibiting the growth of *P. infestans*. The efficacy of bacterial and fungal antagonist found effective as lowest average disease severity (27.89%) was recorded in treatment when *Bacillus subtilis* (B5–0.25%) + *Trichoderma viride* (TV-0.7%) was applied before disease appearance followed by cymoxanil 8% + mancozeb 64%WP (0.3%) at onset of late blight and one more spray of B5 + TV after 7 days.

The different isolates of *Trichoderma* were evaluated against *P. infestans* and found that *Trichoderma* isolates HNA 14 was most effective under both laboratory and field conditions and showed mycoparasitism against *P. infestans* when observed under scanning electron microscope, whereas *T. koningiopsis* and *T. asperellum* were effective against *P. infestans* under both laboratory and field conditions. Rhamnolipid bacterial based formulation (0.25%) was tested under field trials at three different locations for managing late blight of potato. It was observed that the terminal disease severity in rhamnolipid formulation sprayed plot was 45% (against control plot 100%), 47.5% (against control plot 92.5%) and 59.2% (against control plot 76.64%) at Modipuram, Lawar (Meerut) and Jalandhar, respectively.

The some phyllospheric microorganisms *viz.*, yeasts *Sporobolomyces* spp., *Acetenobacter* spp., isolates of *Pseudomonas* spp. and *Bacillus* spp. were reported antagonistic to *P. infestans*. The *Bacillus* sp. inhibited mycelial growth of 7 plant pathogenic fungi *in vitro* and *in vivo* and the same bacterium protected tomato plants against *P. infestans*. A bacterium (*Serratia* sp.), and 4 fungi (*Trichoderma* sp., *Fusarium* sp. and 2 *Penicillium* spp.) were tested against *P. infestans* on tomatoes under field conditions and found that *Penicillium* reduced the lesion area/plant between 8 and 40%.

One hundred twenty two microorganisms isolated from the phyllosphere of potatoes on the development of *P. infestans*, 23 effective microorganisms (spore-forming and non-spore-forming bacteria, yeasts and fungi) were tested in dual cultures and different patterns of inhibition of *P. infestans* were observed. Various naturally occurring microorganisms, i.e., *Trichoderma viride*, *Penicillium virdicatum*, *P. aurantiogiseum*, *Chetomium brasilense*, *Acremonium strictum*, *Myrothecium varrucaria* and *Penicillium aurantiogriseum* showed antagonistic effect against *P. infestans*. The antagonistic activities of *Pseudomonas fluorescens*, *Pseudomonas* sp., *Aspergillus flavus*, *A. niger*, *Penicillium* sp., *T. virens* and *T. harzianum* were tested *in vitro* conditions against *P. infestans*, *Fusarium sp.* and *Rhizoctonia solani*. All bio-agents inhibited the mycelial growth of the pathogens in comparison to control.

The defense enzymes viz., chitinase and β. 1, 3-glucanase activities of *B. subtilis* and *T. harzianum* were well reported against late blight of potato and early and late blight of tomato. Forty-three bacteria were isolated from the phylloplane and rhizosphere of potato and canola plants, evaluated against *P. infestans* causing late blight on potato. It was reported that more than one system (*in vitro* culture media, detached leaves, and whole plants) should be used for selecting and identifying potential of bioagents. A well-known group of microorganism used is the fluorescent *Pseudomonas* which excretes secondary metabolites including antibiotics and biosurfactants that are inhibitory to plant pathogens.

Naturally occurring surface active compounds derived from micro-organisms are called biosurfactants. These are amphiphilic biological compounds produced extra-cellularly as part of the cell membrane by a variety of bacteria, yeast and fungi. Biosurfactants can be used as alternatives to chemical surfactants as their capability of reducing surface and interfacial tension with low toxicity, high specificity and biodegradability make them important for inhibiting pathogens. The best antagonistic activity against *P. infestans* is observed in the genera of *Pseudomonas* and *Bacillus* as they produce wide range of antibiotics and biosurfactants and can be used as alternatives to chemical surfactants. The metabolite of biosurfactant producing bacterium, *P. aeruginosa* has shown high efficacy against *P. infestans* under *in vitro* conditions. Ninety five isolates of bacteria were evaluated for their biosurfactant as well as biocontrol activity against *P. infestans*. It is observed that only 15.8% isolates showed biosurfactant activity and only five isolates were found effective against *P. infestans* for biocontrol properties.

Cultural practices

The cultural practices, includes inoculum free seeds and planting materials, crop and field sanitation and adjustment of crop cultures. Cultural practices classified into three categories: i. Practices, which are usually applied for agriculture purposes not directly connected with crop protection, such as fertilization and irrigation. They may or may not have a positive or a negative side effect on disease incidence or severity, ii. Practices that are used completely for disease control, such as sanitation and flooding and iii. Practices, which are used for both agricultural purposes and for disease control, such as crop rotation, grafting and composting.

Late blight of potato can be managed up to some extent using cultural practices. The infected potato tubers are the primary source of inoculums for causing initial infection of late blight. Besides, areas wherever both mating type (A1 & A2) are co-existed, oospore formation takes place and a possibility to survive longer period in the soil and cause the infection from soil sources also. The oospores as soil-borne inoculums and its significant are determined by formation of oospore in plant tissue and their survival in soil. There is a clear cut correlation between crop rotation and early infections of late blight disease. Generally, infection starts early in fields which are not used for crop rotations. The decline in early infection was most pronounced in fields subjected to crop rotations for three or more years between the potato crops.

It might be a reason that inoculums are less survived in non-crop rotation field than the crop rotated fields. It is clearly indicated that practices of crop rotation is an important aspects for reducing the risk of soil-borne infections of *P. infestans*. The date of potato planting is also useful to avoid the late blight of potato, especially by changing in planting dates. On average, planting in the last 10 days of September resulted in less severe late blight epidemics. Mixed cropping, barrier crops and strip cropping are also helpful for reducing disease severity of potato late blight.

Concept of mixed cropping and barrier crops were investigated for managing/ delaying the spread and build up of late blight in western Uttar Pradesh at Meerut. Results revealed that spreads of the disease were delayed by 7 days by planting resistant cultivars in alternation with susceptible one whereas barrier crop (oat) delayed the spread of disease by 4 days. Strip cropping of potatoes significantly reduced late blight severity in organic production when the crop was planted perpendicular to the wind neighbored by grass clover. Control of contaminated sources such as infected tubers, volunteer plants, waste heaps, disease in neighboring fields and re-growth after haulms destruction can help in management of the disease.

It has been assessed that onset of epidemic can be delayed by 3 to 6 weeks if all primary infection from early potato can be eliminated. It has been shown that during most years late blight epidemics start from infected plants on dumps. Covering of dumps with black plastic sheet throughout the season and preventing seed tubers from becoming infected is an important step to reduce the primary inoculum. Avoiding use of excess nitrogen and use of moderate nitrogen fertilization is often recommended as cultural practices to delay the development of late blight.

Higher dose of phosphorus and potassium has been found to give a higher yield in a late blight year. The selection of suitable cultivars with late blight resistant, well aerated fields, pre-sprouting of tubers and early planting are some of the measures for foliar blight while planting potatoes on large steep ridges, right time of mechanical weeding and harvesting, avoiding rapid shift of harvested tubers or long transports could minimize tuber blight.

Host resistance

Host resistance is the best option for management of late blight of potato and it is eco-friendly in nature. Generally, after a decade, resistant level of the cultivars is being defeated, due to matching of new virulence genes. To find out the source(s) of resistance to late blight in potato was serious concern after Irish famine, during late 19th century. The fact that *P. infestans* originated in Mexico where lots of wild *Solanum* species also grow and co-exist with late blight led to the belief that wild *Solanum* species would possess a fair degree of resistance to balance the *Phytophthora* attack.

In India, selection of late blight resistant genotypes dates back to 1936 when potato germplasm was screened in the field. In subsequent selections, clones of *S. demissum* and *S. antipoveizii* were found immune and later used as parents for late blight resistance breeding. Development of resistant cultivars and exploitation of screening methodology has played an important role in the management of late blight.

CPRI has released varieties having moderate to high degree of resistance to late blight for cultivation both for plains and hills. Some of them are Kufri Giriraj, K. Shailja, K. Himalini and K. Himsona (for hills) and K. Pukhraj, K. Anand, K. Sutlej, K. Badshah, K. Arun, K. Jawahar, K. Garima, K. Chipsona-1, K. Chipsona-2, K. Chipsona-3 and K. Frysona (for plains). Advanced hybrid MS/99–1871 derived from cross PH/F-1045 X MS/82–638 has been released for commercial cultivation under the name Kufri Garima.

Foliage resistance of advanced hybrids tested under laboratory and field conditions did not establish close relationship. The expression of late blight resistance in foliage and tuber were not related. K. Mohan is a new variety with field resistance to late blight reported.

Recently, Payette Russet: a dual-purpose potato cultivar with late blight resistance (both tuber and foliage) and high resistance to potato virus Y released in USA. Somatic hybrids having high degree of resistance to late blight can be used as one of the parent for potato breeding. The somatic hybrids P4, P8 and P10 reported for the introgression of important characters such as high tuber dry matter concentration, resistance to late blight into the cultivated potato via conventional breeding methods for cultivar development in the sub-tropical plains of India.

Forecasting models

The late blight pathogen is highly dependent on the environmental factors like temperature, relative humidity and leaf wetness etc. for causing late blight disease. Therefore, various forecasting model had been developed for forecasting late blight disease. Initially, Van Everdingen evolved 'Dutch rules' for predicting the initial occurrence of late blight and for scheduling fungicide applications under Holland condition. Subsequently, Beaumont's period; Irish rules, moving day concept; severity value accumulation; negative prognosis and mathematical based models were developed worldwide.

Large number of forecasting systems like BLITECAST, SIMCAST, ProPhy, PROGEB, PhytoPre, NegFry, Web-Blight, Plant Plus, PhytoPRE + 2000, China Blight, Bio-PhytoPre etc. have been developed for different regions of the world. International Potato Centre has linked two disease forecasting models, Blitecast and Simcast to climate database in a Geographical Information System (GIS) to estimate global severity of potato late blight. Using GIS database, they suggested that an increased access to host resistance and fungicides in developing countries could have a strong economic impact on potato production.

A web-based Decision Support System (DSS) was developed for management of potato and tomato late blight which links various models into a system that enables prediction of disease dynamics based on weather conditions, crop information, and management strategies. Growers identify the location of their production unit of interest and the system automatically obtains observed weather data from the nearest available weather station, and location-specific forecast weather data from the National Weather Service – National Digital Forecast Database.

Recently a new forecasting model BLITE-SVR developed for prediction of first appearance of late blight of potato. A total of 13 kinds of weather data had been utilized for development of this model and performance of BLITE-SVR compared with the conventional moving-average method as well as through pace regression and linear regression. The accuracy of prediction was 64.3% by BLITE-SVR, with 42.9% by the conventional moving-average method, 42.9% by pace regression and 35.7% by linear regression for first appearance of late blight of potato.

In Indian scenarios, a forecasting model has been developed for Darjeeling hills utilizing 12 years rainfall data on the concept of Cook's moving graph and Hyre's. Another forecasting model had been developed using daily weather data (temperature, rainfall and RH) for actual appearance of late blight for Shimla, Shillong and Ootacamund.

The computerized forecasting model 'JHULSACAST' developed for western UP for both the rainy and non-rainy conditions and it is being utilized for forecasting of first appearance of late blight in the regions and large scale of farmers are benefited by timely adopting control measures.

The wireless sensor network was used for validation of 'JHULSACAST' with other forecasting late blight models in western Uttar Pradesh using human participatory sensing approach.

It was observed that the 'JHULSACAST' has been found to be significantly accurate than the Ullrich, Fry, Winsteland Wallin models for the Hapur region of Uttar Pradesh, India. JHULSACAST model template was used for calibration for development of forecasting models for Punjab, Tarai region of Uttarakhand and plains of West Bengal.

A decision support system also developed for assisting in management of late blight by ICAR-CPRI, which includes three modules i.e. i) decision rules for forecasting first appearance of late blight in plains during rainy and non-rainy years based on temperature, relative humidity, and rainfall data, ii) decision rules for need based application of fungicides, and iii) regression models for yield loss assessment. All these modules have been combined and a web based decision support system for western Uttar Pradesh has been developed and hosted on ICAR-CPRI server.

The yield loss assessment model was developed using two parameters *i.e.* per cent yield loss as a dependent variable and AUDPC as an independent variable. Twenty five linear and non-linear regression lines were fitted with three years data and amongst best non-linear reciprocal hyperbola regression line, which has $R^2 = 0.84$ was selected.

Further, this model was validated and results revealed that the deviation from 0.5 to 13.70% in 2010–2011, 1.16 to 9.69% in 2011–2012 and -3.01 to 9.23% in 2012–13 between actual and predicted yield loss. Recently, INDO-BLIGHTCAST- a web based Pan-India model for forecasting potato late blight which is an improvement over JHULSACAST has been developed. It predicts late blight appearance using daily mean temperature and relative humidity data available with meteorological stations and does not require hourly weather data, not region/location specific and can be used across the country without any calibration. An algorithm to determine the severity of potato late blight was developed using image processing techniques and neural network. The proposed system takes images of a group of potato leaves with complex background as input which are captured under uncontrolled environment. It could further modified for spray of fungicides based on disease severity. Thus, the disease forecasting model is not only forecast for initial appearance of late blight but also assist in managing the late blight with proper spray schedules.

Fungicides: Good Preventives when Growing Potatoes

A high-quality potato yield is often dependent upon the use of crop protection products. Fungicides are especially important, since many potato diseases have damaging symptoms that don't manifest right away.

Pink rot, a soilborne pathogen that thrives in saturated soils, infects plants through the roots and can go undetected until harvest time. Late blight is another infection that can ruin potato crops - in favorable conditions, late blight expands rapidly, turning dying tissues from brown to purplish-black. With so many diseases that have unpredictable symptoms, taking preventative action could be the difference between distress and success.

PULSE CROPS

Pulse crops are members of the family Leguminosae or Fabaceae. The term "pulse crops" generally refers to those plant species harvested primarily for dry seed that is used as both human food and animal feed.

Pulse crops are considered as environmentally friendly due to their ability to form symbiotic associations with nitrogen-fixing *Rhizobium* bacteria in their taproot systems. The fixed nitrogen in the soil is available to the subsequent crop. Also, inclusion of pulse crops in the crop sequence reduces disease incidence in crops that follow a pulse crop. The fruit of a pulse crop is a pod containing one to several seeds, which are high in protein, carbohydrates, fibre, minerals and vitamins (folate). Consumption of pulses as part of a regular diet may offer numerous health benefits including reduced risk of diabetes and heart diseases.

Pulses are annual leguminous crops yielding between one and 12 grains or seeds of variable size, shape and colour within a pod, used for both food and feed. The term "pulses" is limited to crops harvested solely for dry grain, thereby excluding crops harvested green for food, which are classified as vegetable crops, as well as those crops used mainly for oil extraction and leguminous crops that are used exclusively for sowing purposes.

Besides serving as an important source of protein for a large portion of the global population, pulses contribute to healthy soils and climate change mitigation through their nitrogen-fixing properties. Bengal Gram (Desi Chick Pea / Desi Chana), Pigeon Peas (Arhar / Toor / Red Gram), Green Beans (Moong Beans), Chick Peas (Kabuli Chana), Black Matpe (Urad / Mah / Black Gram), Red Kidney Beans (Rajma), Black Eyed Peas (Lobiya), Lentils (Masoor), White Peas (Matar) are major pulses grown and consumed in India.

History

"Pulse" is a derivation from the Latin words *puls* or *pultis* meaning "thick soup". Pulse crops are small but important members of the legume family, which contains over 1,800 different species. Pulse crops are the seeds of legumes that are used as food, and include peas, beans, lentils, chickpeas and faba beans. Evidence of cultivation of lentils has been found in the Egyptian pyramids and dry pea seeds have been discovered in a village in Switzerland dating back to the Stone Age. Archaeological evidence suggests that peas were grown in the eastern Mediterranean and Mesopotamia regions at least 5,000 years ago and in Britain as early as the 11th century.

Pulses are an important source of protein, especially in developing countries. Pulses provide about 10% of the total dietary protein consumed in the world and have about twice the protein content of most cereal grains. Pea crops were a leading production crop in eastern Canada at the turn of the century, with an average of 720,000 acres (288,000 hectares) grown each year from 1883 to 1902. Production in eastern Canada gradually declined; by 1970, only 82,000 acres were grown in all of Canada with about 70% of that production in Manitoba. Pulses did not play a significant commercial or economic role in Western Canada until the 1970s, when the wheat glut encouraged farmers to diversify into cash crops such as rapeseed (canola), lentils, peas and other specialty crops.

Classification of Pulse Crops.

Grams

1. Red gram/Pigeon pea : Cajanuscajan
2. Bengal gram/

Chick pea : Cicerarietinum

1. Black gram/Urd : Phaseolusmungo/Vignamungo
2. Green gram/Mung : Phaseolusaureus/Vignaradiata
3. Horse gram : Macrotylomauniflorum

Beans

1. Soybean : Glysine max
2. Field bean/Avare : Dolichos lablab
3. French bean : Phaseolusvulgare
4. Lima bean/

Double bean: Phseoluslunatus

1. Winged bean : Psophocorpustetragonolobus
2. Cluster bean : Cyamopsistetragonoloba
3. Moth bean : Vignaaconitifolia

Peas

1. Pea : Pisumsativum
2. Cow pea : Vignaunguiculata/V. sinensis

There are a limited number of foliar fungicide active ingredients registered for lupin diseases in WA, in either narrow-leafed or albus lupins. Products are currently registered for anthracnose, botrytis grey mould and sclerotinia.

Fungicide registrations in other pulses

Foliar fungicides are registered for the following pulse diseases:

- Chickpea: Ascochyta blight, Botrytis grey mould;
- Field pea: Downy mildew, Botrytis grey mould, Blackspot and Powdery mildew;
- Faba bean: Ascochyta blight, Cercospora, Chocolate spot and Rust;
- Lentil: Ascochyta blight, Botrytis grey mould;
- Vetch: Ascochyta blight, Botrytis grey mould and Rust.

Registrations are shown in the pulse document available under 'Documents' on the right.

How to use fungicides effectively

Effective fungicide use depends on the following areas:

Correct diagnosis of a fungal disease

Fungicides are effective only on diseases that are caused by fungi. Sometimes plant symptoms resemble those caused by pathogens but they are actually caused by non-pathogens, such as nutrient deficiencies or toxicities and adverse weather conditions. A disease diagnosis service is provided by DDLS - Plant pathology services. PestFax provides useful up to date information on diseases present throughout the wheatbelt each season.

Appropriate selection of a fungicide from available registered products

Once you have correctly identified the disease, refer to the registrations document available under 'Documents' on the right for registered active ingredients in WA. For details on which products or trade names contain a specific active ingredient search the online PubCRIS database or Infopest (subscription required) or contact your agronomist or chemical provider. Foliar fungicides are currently only registered for managing anthracnose, botrytis grey mould and sclerotinia stem rot. Many of the lupin registrations are APVMA permits that have specific expiry dates.

It is vital that product labels and permits are read before application to identify: that the chemical is registered for use on the crop and for the spectrum of diseases present, the withholding period and any specific application instructions or directions for use. Be careful of tank mixes with other products. In some instances fungicide sprays need to be mixed with an adjuvant.

Correct timing of fungicide application

Most of the information on optimal application timing is available on fungicide product labels. Fungicides are better applied before or in the initial stages of a disease outbreak so being aware of your disease risk and routine crop monitoring is needed for timely application of fungicide. Follow-up fungicide application may be needed to extend protection in some situations. Refer to specific DPIRD disease management pages on this site for recommended application timings, such as Lupin foliar diseases: diagnosis and managment, external link available on the right.

Reduce the risk of fungicide resistance developing

Fungicide resistance develops from misuse of fungicides and poor disease management practices. There are currently barley powdery mildew and net blotch populations resistant to some fungicides in WA so this is a very real threat that we are facing. You can reduce the risk of fungicide resistance developing by growing varieties less susceptible to disease (where available), supported by integrated disease management in conjunction with strategic and responsible use of fungicides (eg rotating and mixing fungicides with different modes of action).

Pulse Seed Treatments

Seed treatments and foliar fungicides are part of an overall strategy to reduce losses from disease in pulse crops. The best results are achieved when fungicides are combined with clean seed, crop rotation, paddock selection, resistant varieties, and crop hygiene. Even though disease is reduced on resistant varieties, seed treatments and foliar fungicides can enhance their performance. Check that the fungicide registration or permit is current for the crop and disease targeted. Comply with all grazing withholding periods and any export slaughter intervals.

Treating pulse seed with a fungicide reduces the establishment of seed-borne diseases in crops and also protects seed from infection by soil-borne fungi. Some fungicide seed dressings also protect seedlings from external airborne infection for the first few weeks. This is important in reducing the subsequent establishment and spread of disease within crops. Seed treatments provide effective control for a maximum of four to six weeks after sowing, but do not provide absolute control.

Application

It is important for seed treatments to be evenly distributed on seed to ensure each seed gets an effective dose. This is enhanced for flowable seed treatments by dilution with water (refer to the label). Secondary mixing of treated seed through an auger assists to obtain even seed coverage. Correct calibration of the applicator and a consistent seed flow are critical for the recommended rate of seed treatment to be applied.

Applying fungicides and rhizobium inoculant to seed.

Some fungicides are toxic to rhizobium, and should not be mixed together before application to seed. Read the labels for compatibilities. Ideally, seed should be treated with fungicide and then, in a separate operation, inoculated with rhizobium just before sowing. Sowing should occur immediately after rhizobium has been applied, particularly in acid soils. Granular or liquid injection of inoculum infurrow eliminates the contact between seed treatments and rhizobia.

Foliar Fungicides

Foliar fungicides are necessary for controlling some destructive pulse diseases. They are most effective when applied before or at the first sign of disease, or immediately prior to weather conditions favourable for disease development. Do not wait until the disease is established i.e. apply ahead of rain fronts, as pulse fungicides protect rather than cure. Fungicides provide protection for 10-20 days. The duration of protection varies with the product used, how rapidly the plants are growing, and the rainfall experienced. Any new growth after spraying is not protected. If disease persists, additional sprays will be required and should be applied prior to rain. Waiting until after a rain event allows infection to occur.

Application

Uniform coverage of the crop foliage is important for prevention of disease. This is best achieved by using high water rates (preferably 100 L/ha by ground and 30 L/ha by air) with water pH not exceeding pH7, nozzles with a fine or extra fine droplet spectrum and an operating pressure of 400 kPa is suggested. Application onto a wet plant (heavy dew) can assist in coverage of the product.

Lupin

Brown leaf spot/Pleiochaeta root rot (Pleiochaeta setosa)

Brown leaf spot can kill seedlings by defoliation up to the 4-leaf stage. Seed dressing can reduce leaf drop by 50%. Seed treatments containing either ®iprodione (eg Rovral®, Drover®, Xlflo , ipridione® products) or procymidone (Sumisclex®, Spiral , ®Lupinflo) will reduce the transfer of seed-borne disease to seedlings. Remember if applying a fungicide seed treatment and seed-applied rhizobium, precautions are necessary to maximise the survival of rhizobium inoculum and nodulation. Pleiochaeta root rot is only suppressed by seed treatment, but not completely controlled.

Foliar fungicides will help manage the lentil diseases Ascochyta blight (AB) and Botrytis grey mould (BGM), but when to commence a fungicide program and how often to spray, depends on the varietal resistance, weather conditions and the potential yield of the crop.

Pulse Australia have a number of Minor Use Permits from APVMA in place for 2019 to help growers with disease control and these are shown in the table below along with all the current registered products. It is particularly important to observe all the conditions set out in these permits as lentil are grown for the human consumption market, here and overseas, and market access depends on the strict delivery standards and residue limits. There is no secondary market for lentil if these standards are not met.

Fungal disease control is based on the use of integrated disease management to minimise the injury to crops from plant pathogens. Efficient use of foliar fungicides is based upon the protection of plants rather than curing existing infections.

There are three critical periods for fungicide spraying decisions on lentil crops for disease.

- First critical period is at 10-14 weeks after emergence (WAE), shortly prior to canopy closure. Early application of fungicide is critical in restricting the early development and spread of BGM. In susceptible varieties, or in districts prone to BGM epidemics, apply a fungicide, irrespective of BGM being present. An application at this stage is the final chance for spray penetration deep into the crop canopy to protect stems.

- Second critical period is at mid flowering/early pod fill (14-16 WAE). If either BGM or AB is present, or weather is conducive to disease, apply a fungicide. The type of fungicide used will be dependent on the variety. Mixtures of foliar fungicides may be required to give control for both diseases in some susceptible lentil varieties.
- Third critical period is at the end of flowering/mid pod fill (16-18 WAE). This is the final growth stage where all pods are formed and protection against AB infection ensures good seed quality is achieved.

Foliar fungicides remain effective for approximately 10-14 days under ideal conditions. Keep in mind that all new growth after spraying is unprotected. Timing of fungicide applications is critical. An application in advance of a rain front provides maximum protection. Delaying application until after a rain front reduces efficacy significantly as diseases can be rapidly spread following rain. Close monitoring for early symptoms will give greater opportunities to minimise disease establishment and spread.

The need for repeated fungicide applications depends on the growth stage of the crop, the time since the last fungicide application and the likelihood of further conditions favouring disease development. Unprotected crops may be severely affected by disease impacting on yield and grain quality particularly in wet springs.

Management of foliar diseases of food legumes

The main emphasis in research and development to combat food legume diseases is on host resistance and chemical control where ever applicable, and quite often these components of disease management were practiced in isolation to each other. Recently a shift in scientific thinking and practice in the management of grain legume diseases has been seen and greater emphasis was on identifying, evaluating, and integrating location specific components of integrated disease management (IDM). In general IDM has followed the principles of IPM. The location specific IDM of food legumes is primarily based on host plant resistance (HPR) or genetic resistance; additionally other components of diseases management. In some environments, IDM may require a single component used alone (usually HPR) or in combination with one other component (such as fungicide seed treatment) to adequately combat diseases of food legumes. The components of IDM employed in the production of food legumes are listed as follows:

Host plant resistance (HPR),

Disease modeling (prediction) for avoidance of high risk or disease pressure,

Chemical sprays (fungicides, pesticides),

Biological control, and

Cultural (agronomic) practices (sowing dates, plant population etc.).

Cool season legumes

Chickpea (Cicer arietinum L.)

Major foliar diseases of potential economic importance in chickpea are ascochyta blight (AB) and botrytis grey mold (BGM). These diseases have been comprehensively reviewed by several workers in the past have discussed advances in etiology, biology and management of diseases of food legumes and Singh et al. (2007) have discussed chickpea diseases and their management in detail.

Ascochyta blight infection and disease progression occur from 50 to 25 °C with an optimum temperature of 16-20 °C, and a minimum of 6 h leaf wetness. Disease severity increases with the increase in relative humidity. Cloudiness and prolonged wet weather favour rapid development and spread of both the diseases. The pathogen survives on infected or contaminated seeds, infected chickpea debris which causes AB, produces both rain splashed conidia and wind blown ascospores. The existence of 2-12 races of A. rabiei has been proposed by several researchers. Researchers grouped 14 isolates of *A. rabiei* into eight pathotypes at >0.5 similarity coefficient.

Infection of BGM occurs from 15-25 °C with an optimum temperature of 20-25 °C. High moisture and high relative humidity are congenial for BGM development. Under such conditions there is abundant sporulation of the fungus B. cinerea on dead plant parts, particularly on flowers and pods. The pathogen B. cinerea is reported to have extreme diversity and adaptability to a wide range of environmental conditions. Existence of 4 - 5 pathotypes of B. cinerea has been reported from northern India.

Adoption of IDM practices is essential for economical and effective control of AB and BGM. In several studies conducted in different chickpea growing areas of the world, several sources of resistance to AB were identified. Furthermore, development of AB resistant genotypes has made it possible to sow the crop during winter in the Mediterranean region thereby doubling the chickpea production potential. On the other hand, an adequate level of genetic resistance to BGM is not available in the cultivated genotypes and fungicides become ineffective under the high disease pressure. Hence, IDM using the available management options is essential to successfully manage the disease and mitigate yield losses. Moderate level of HPR can be combined with other cultural practices and/or application of minimum dosage of fungicides for control of AB and BGM.

A combination of a moderately resistant variety and 2 sprays of chlorothalonil, one during the seedling stage and another at the early podding stage, provided the most economical field control of AB in Syria and Australia. In collaboration with the Syrian national program, ICARDA has developed an IDM package for AB management. The components of this package include use of tolerant cultivars adapted to early sowing, seed dressing with fungicides, and single foliar application of chlorothalonil at seedling or early vegetative growth stages. This package resulted in higher chickpea yields compared with the traditional spring plantings using a local variety without seed dressing or fungicide spray.

Agronomic and cultural management of BGM has been demonstrated in India, Bangladesh and Nepal. In Bangladesh, an IDM package, comprising cultivation of the BGM tolerant genotype "Avarodhi", soil application of diammonium phosphate, wider row spacing , seed treatment with carbendazim + thiram (2g/kg seed), and need based foliar application of carbendazim, has been devised.

The location-specific recommended IDM practices for AB include: (a) use of pathogen-free seed, (b) seed treatment with fungicides, (c) practice of crop rotation, (d) deep ploughing of chickpea fields to bury infested debris, (e) use of disease-resistant genotypes, and (f) strategic application of foliar fungicides.

Lentil (Lens culinaris Medik.)

Ascochyta blight, rust and powdery mildew are economically important foliar diseases of lentil. Ascochyta blight is caused by A. lentis produces conidia, in a flask-shaped fruiting body (pycnidium), which are spread by rain splash. The teleomorph, Didymella lentis, produces ascospores which can be wind dispersed large distances. The pathogen is both stubble and seed borne. Progress of foliar blight is rapid and epidemic levels can be reached under cool and wet weather conditions as a consequent of spores being disseminated by rain splashes. It survives for more than 3 years in infected pods and seeds at 4-5 °C or under shelter outdoors, and 1-5 years at the soil surface. IDM of Ascochyta blight of lentil includes use of resistant cultivars, use of disease free seed, crop rotation, seed treatment and foliar spray. Application of the fungicides benomyl, carbendazim, thiabendazole, chlorothalonil, prothioconazole and strobilurin, are useful for managing Ascochyta blight epidemics.

Lentil rust is caused by Uromyces vicia- fabae (Pers.) de Bary. It is an autoecious fungus; completing its life cycle on lentil. The disease occurs during the flowering/early podding stage as aecia, which may develop into secondary aecia or uredia. The resulting aeciospores and uredospores lead to a further disease spread in the crop season. Uredia rapidly appear a little late in the crop season followed by development of telia. The fungus survives in summer as teliospores. High humidity and cloudy weather with temperatures of 20-22 °C favor disease development. The plants give dark brown or blackish appearance visible as patches in the field.

Integrated management of rust includes control of volunteer plants over the summer and removal of infected lentil debris. It is advisable to use clean seeds without rust contaminations, and to treat the seed with a suitable fungicide such as diclobutrazole. Preventive fungicide sprays of mancozeb at early disease development stage have been recommended. The use of host plant resistance is the best means of rust management.

Several rust resistant cultivars have been released in different countries, with resistance originally identified at ICARDA, Syria and in India. There is no clear evidence for the existence of races. Both complete and partial resistance exists and monogenic resistance has been reported. Field resistance of lentil to rust is governed either by a single dominant gene, by two independently inherited dominant genes, by two complementary genes or by three independently inherited genes.

Powdery mildew another important disease of lentil has been reported from many countries such as Cyprus, Ethiopia, India, Siberia, Sudan, Syria, Tanzania and the former USSR. In North America, powdery mildew is often observed after lentil started flowering. The disease poses a serious problem on breeding material in plastic or glass houses in both India and Syria, and in India it is also recorded in off-season nurseries in Trans Himalayan regions, like Lahaul Spiti and Sangla etc., but it is rarely seen in the field during the cropping season. A fine powdery, white growth of conidia and mycelium initiates as small spots and rapidly spreads to cover the entire surface of leaves, stems and pods. Later, the leaflets become dry and curled, and are shed prematurely, causing considerable reduction in yield and seed quality. The seeds from infected plants remain small and shrivelled.

Powdery mildew of lentil is caused by the ectoparasites *Erysiphe pisi* DC., and *E. polygoni* DC. and the endoparasite *Leveillula taurica* (Lév.) Arnaud. Recent evidence showed that *E. trifolii* also infects lentil. The anamorph stage is responsible for spread of the disease. The teleomorph stage has been reported to occur in India and Sudan. Moderately high temperatures and moderate relative humidity favour the disease development. Many lentil genotypes are reported resistant to powdery mildew, and should be planted whenever possible. Foliar sprays with fungicides benomyl, tridemorph, aqueous sulfur, karathane (dinocap), calixin or sulfex as well as certain insecticides (Quinalphos, Tnazophos, Phoxim) applied at 10-15 days interval are effective in suppressing mildew growth.

Faba bean (Vicia faba L.)

Ascochyta blight, chocolate leaf spot and rust are the major diseases of faba bean. Ascochyta blight also known as leaf, stem, and pod spot is caused by *Ascochyta fabae*. Both stages are described by researchers. The fungus is highly specialized to faba bean and is highly variable in culture. Researchers identified seven genes of resistance and resistance was either monogenic or oligogenic. Limited information is available about the biology of *A. fabae*. More work needs to be done to standardize methods and differential cultivars to establish the A. fabae races.

Chocolate leaf spot is caused by both *Botrytis cinerea* Pers. ex Pers. and *Botrytis fabae* Sard. A teleomorph of B. fabae, *Botryotinia fabae*, was described by researchers. *B. cinerea* is a parasite and saprophyte on a wide range of host plants, whereas *B. fabae* is specialized for the invasion and colonization of *Vicia* spp. especially *V. faba*. The existence of races of B. faba has been proposed on the basis of reaction to differentials in Mediterranean countries. More work needs to be done to understand variability in *B. fabae.*

Another disease of faba bean, brown rust is caused by the fungus *Uromyces viciae fabae* (Pra.) Schroter. The pathogen predominantly produces aeciospores and uredospores on leaves, and teliospores are formed in large black sori on stems and petioles. While earlier studies mostly focused on cytological aspects, later studies were concerned with biochemical and molecular characteristics. Despite the fact that there is still no stable transformation system available for any obligate biotroph, recent molecular analyses have provided new insights into this highly sophisticated interaction of a fungus with its host.

IDM strategy for controlling foliar diseases of faba bean includes the use of disease free seed, avoid sowing too early to minimise disease, follow 3-4 year crop rotation and select the tolerant variety to the main disease risk in particular region. Thorough and regular crop monitoring is essential if the strategic spray program is to be successful. The timing of fungicide applications depends on the disease level observed, the time since the previous application, and the likelihood of rainfall and other conditions conducive to infection and spread of chocolate spot. Carbendazim is best used when high chocolate spot pressure occurs or when rapid plant growth produces large amounts of unprotected foliage, particularly from mid-flowering onwards. Cholorothalonil and mancozeb are best to be used earlier if rust or Ascochyta is a problem.

Field pea (Pisum sativum L.)

Peas are adversely affected by serious fungal, bacterial and viral diseases such as: powdery mildew (Erysiphe pisi Syd. (syn. E. polygoni DC), Ascochyta blight or black spot (Ascochyta pisi/A. pinodes), downy mildew (Peronospora pisi), bacterial blight (Pseudomonas pisi), Pea early browning virus (PEBV), Pea enation mosaic virus (PEMV) and Pea mosaic virus (PMV). Researchers has comprehensively described pea diseases.

Powdery mildew occurs all over the world and can cause severe damage in areas where pea is cultivated. Powdery mildew of pea caused by Erysiphe pisi, is a serious disease both in the field and in the greenhouse. All aboveground parts of plants are susceptible to powdery mildew. Pod infection may discolor seeds to a gray brown color. The powdery look of the disease is caused by the profuse production of conidia on the upper leaf surface. Management of powdery mildews of grain legumes is through use of resistant cultivars, especially in late sown crops, which are likely to experience high disease pressure. Resistance in pea is conditioned by two recessive genes (er-1 and er-2) along with two or more modifying genes.

Resistance in cultivars homozygous for er-2 is expressed mostly in leaves and this resistance can be rendered ineffective under high disease pressure. The disease is often less severe in areas where overhead irrigation is applied regularly because long periods of free water on host leaves reduce conidium viability and wash conidia from host tissue.

Other control measures include fungicide sprays of sulfur and/or demethylation inhibitors such as cyproconazole, fenarimol and triadimenol. Fungicide spray should be applied at least two weeks before harvest to avoid residue on peas.

Ascochyta blight or black spot is the most common and most damaging disease of field pea in southern Australia. Worldwide, the disease is recognized as being caused by any one, or combinations, of three fungi; *Mycosphaerella pinodes*, *Phoma medicaginis* var. *pinodella* and *Ascochyta pisi*. All three frequently occur together hence the disease is generally referred to as the ascochyta complex of peas. *Mycosphaerella pinodes* causes the most damage to pea crops in Western Australia, Washington, USA and is the principal pathogen involved in nearly all occurrences of blackspot. *Mycosphaerella pinodes* survives on pea stubble for more than 3 years producing viable ascospores during each growing season. Ascospores are released from pseudothecia on the stubble following rain events of as little as 0.2 mm. The airborne ascospores can infect crops several kilometres away.

IDM includes use of moderately resistant varieties, disease free seed, crop rotation, delay in the sowing, disease forecast model that predicts the time of onset, and progression of ascospores maturity and spread of spores from the source of infection and need based foliar and in-furrow applications of fungicides in conjunction with other agronomic practices.

Warm season legumes

Pigeonpea (Cajanus cajan (L.) Millsp.

Sterility mosaic disease (SMD) is the most important foliar disease of pigeonpea in India and Nepal. Etiology of sterility mosaic is unknown despite of numerous attempts during the past 20 years. Researchers reported a tenui virus of asymmetric morphology as the cause of SM disease and retained the name of virus as Pigeonpea sterility mosaic virus (PPSMV). PPSMV is flexous, branched filaments measuring 3-8nm in diameter. The SM causal agent is not transmitted through sap or seed. It is transmitted by the eriophyid mite vector.

Information on physiological specialization of SM is limited and is based on symptoms observed on the inoculated plants using the mite vector. Recently PCR based techniques are being developed for distinction of different species of eriophyid mites.

HPR is the most reliable and sustainable method for the management of SMD. Considerable progress has been made in identifying resistance sources and developing resistant cultivars to the disease. Some attention has also been paid to cultural and chemical control of sterility mosaic.

Cultural practices include: Destroy ratooned pigeonpea, uproot and destroy infected plants at the initial stage of disease development, crop rotation to reduce inoculum levels and vector populations, chemical control as seed treatment with 25% carbofuran or 10% aldicarb (3g kg^{-1} seed) and spraying acaricides or insecticides like karathane, metasystox to control the mite vector in the early stages of plant growth.

Mungbean (Vigna radiata (L.) Wilczek) and blackgram (Vigna mungo (L.) Hepper

Mungbean (green gram) and urdbean (black gram) are widely cultivated in many Asian countries in different seasons. Three diseases (yellow mosaic, Cercospora leaf spot and powdery mildew) that attack both the pulses are considered economically important. Yellow mosaic caused by mungbean yellow mosaic virus (MYMV) is the most serious limiting factor in mungbean and urdbean cultivation. The pathogen is transmitted by the white fly *Bemisia taba*ci Genn.

Cultivation of resistant varieties, manipulation in sowing dates, inter/ mixed cropping of mungbean and urdbean with non-host crops like sorghum, pearl millet and maize and application of systemic insecticides such as aldicarb, disyston and foliar application of metasystox has been found effective in controlling the disease by reducing vector control.

Cercospora leaf spot caused by *Cercospora cruent*a and *C. canescens* causes severe leaf spotting and defoliation at the time of flowering and pod formation. Involvement of different species in causing cercospora leaf spot complicates characterization of species. Researchers reported yield losses to the tune of 50% in severely diseased field. Since there is low level of resistance to cercospora leaf spot, the cultural practices and chemical control play an important role in its management.

Cultural practices such as field sanitation, crop rotation, destruction of infected crop debris, and avoiding collateral hosts in the vicinity of the crop may help in reducing the incidence. Mancozeb, carbendazim, copper oxychloride and benomyl are reported to reduce disease incidence considerably.

Powdery mildew caused by Erysiphe polygoni DC, is a problem in cool dry weather. Pathogen is obligate parasite and has wide host range. Limited information is available on the etiology and biology of E. polygoni on Vigna mungo and Vigna radiata. Many resistant sources are available against powdery mildew.

Also its incidence can be reduced by adjusting the date of sowing with wider spacing. Chemical control with fungicides karathane, calixin, bavistin, benlate, topsin M, sulfur dust etc. has been found effective to control the disease under field conditions.

Cowpea (Vigna ungiculata (L.) Walp.)

Cowpea is the most important legume vegetable grown in India and to a limited extent in Nepal and Bangladesh. Cercospora leaf spot, cowpea golden mosaic and cowpea aphid-borne mosaic are of potential economic importance. Limited information is available about the distribution, etiology and biology of these diseases. Cercospora leaf spot caused by *Cercospora canescen*s and *Pseudocercospora cruenta* have been observed in all the cowpea growing areas.

Despite the fact that cercospora leaf spot develops late in the season, disease spread is often rapid and premature defoliation can be severe. The disease can be controlled by combining resistant varieties and spray of fungicides such as benomyl and captafol post flowering.

SOYBEAN

Soybean (*Glycine max* (L.) Merr.) is a species of legume native of the Eastern Asia, widely grown for its edible bean, which has numerous uses. For many centuries, it was the main source of food for the people of the Far East (China, Japan, India). The ability of soybean for the food and non-food uses has been recognized for many years. Among available protein sources of plant origin, soybean protein is the closest to the optimum dietary essential amino profiles for human and animal nutrition.

Therefore, soybean production, consumer acceptance and consumption in non-traditional regions of the world are on the rise. Increasing world population, constant need for animal feed and numerous different soybean products contribute to the great demands for soybean seed in world market. Nowadays, soybean is the leading oil and protein crop, whose seed is used as a source of edible oil and protein for both human and livestock consumption and for various industrial purposes. Due to consumer's large interest in healthy food, soybean is grown more and more frequently under the organic system in which it produces satisfactory yield.

Domestication and history

Soybeans have historically been grown as food for millions of people and as an ingredient for livestock feed. Through research and product development, there are many industrial and food use products that use soybeans or soybean-derived products as one of the main ingredients. Since soybeans are composed of protein and oil, soybeans are available for many uses, as processors can separate soy's components into soy protein (meal) and soy oil. Soy protein, which makes up 80 percent of the soybean primarily serves as animal feed and is incorporated in many human foods. The United States food industry uses more than 83 percent of soybean oil for cooking baking and frying. The remaining soybean oil is used in industrial applications such as adhesives, printing inks, plastics and specialty products.

The origins of the soybean plant are obscure, but many botanists believe it was first domesticated in central China as early as 7000 BCE. An ancient crop, the soybean has been used in China, Japan, and Korea for thousands of years as a food and a component of medicines. Soybeans were introduced into the United States in 1804 and became particularly important in the South and Midwest in the mid-20th century. Brazil and Argentina are also major producers.

The environment plays a crucial role in soybean growth and development, and a major study found that an estimated 3.1% soybean yield loss can be expected for each 1°C rise in global mean temperature. However, this estimate does not take into account yield losses due to pathogens, which are also affected by the environment. All three aspects of the plant disease triangle must be present for plant disease to occur: a virulent pathogen, a susceptible host, and a conducive environment (Fig. 17).

In each section below, we touch on one of the greatest threats to modern and future agriculture; the shift in global climate patterns and the increased frequency of extreme weather events. Over the next 30 yr, temperatures are predicted to increase 1.5–2°C (2.7–3.6°F) and precipitation is expected to increase in the United States, with many locations already experiencing dramatic changes in weather patterns.

Predictive modeling for weather systems in the United States estimates that the frequency of convective systems, or very large thunderstorms, will triple by the end of the century, and the total volume of precipitation will increase by 30–80%. Extreme weather events in 2019 led to heavy rainfall, severe flooding, and approximately 8 million hectares (20 million acres) filed for 'prevented plant' insurance coverage, an economic burden of nearly $4 billion. From 1958 to 2012, the Upper Midwest region of the United States has experienced a 37% increase in precipitation, with trends continuing upwards. The impacts these types of changes will have on plant diseases are challenging to predict due to the unique disease cycle of each pathogen, though weather changes are predicted to increase the risk of diseases in the coming decades.

With concerns over changing climate in the 21st century, the environmental component of the disease triangle may also be changing the frequency and severity of disease outbreaks in some regions. In order to prevent a disease, farmers must know if they are at risk for diseases caused by pathogens present in their region, and if their environment is conducive for other pathogens that could enter the area and cause disease. The most significant threats to U.S. soybean production since 2000 have been documented, and strategies to manage the major diseases in the United States are covered in this review.

Additional soybean diseases that have major effects on soybean production in other countries, such as soybean rust and anthracnose, may also become more significant threats in the United States should climate trends continue as projected. Through this review, we aim to provide: 1) an overview of the impact and disease cycle for common soybean diseases in the United States, 2) a comprehensive review of the state of knowledge regarding current management strategies for each soybean disease, and 3) insights into the epidemiology of each pathogen and an indication of the likelihood of outbreaks and spread of major pathogens to new regions based on changing climate trends.

Seed quality

The importance of soybean stems from the quality of its seed. It contains about 35-50% proteins and 18-24% oils, depending of the variety and growing conditions. Soybean has the highest crude protein content and a balanced amino acid profile. Hence, soybeans have become a top choice for protein and as such, they are strategically traded around the world. The main protease inhibitors in soybean – Kunitz trypsin inhibitor and the Bowman-Birk inhibitor constitute the main anti-nutritional factors of soybean. Trypsin inhibitors contribute to reduce digestibility of seed protein.

Nutritional and functional properties of the soya oils are determined by their fatty acid composition, the distribution pattern of the fatty acids within the triacylglycerols (TAGs) and the total content and composition of natural antioxidants. Soybean is considered to be the most abundant natural source of isoflavones in the human and animal diet. Isoflavones are the main components of flavonoids and the most common form of phytoestrogens. Totally dry soybean seed contains 33% DW of carbohydrates on average, of which 16.6% DW is soluble sugars. The significance of the soluble sugar profile is in its effect on quality, digestibility and nutritional value of soybean for food and feed. Soybean seed also contains vitamins (A, B, D and E), minerals (calcium, iron, and potassium) and phospholipids, such as lecithin.

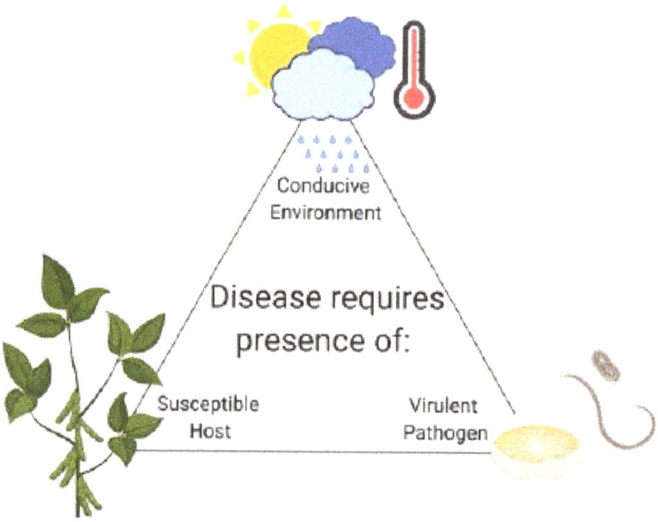

Fig. 17

Utilization

Soybean is used both for oil production and protein processing. Most of the produced soybean in the world is used for animal nutrition. Soybean meal is the highest quality plant protein nutrient that can serve as the basic source of protein for animal, especially all categories of pigs and poultry, then fish, fattening cattle, and is important in the diet of high-milking cows and sports horses. The use of raw soybean in the diet does not give satisfactory results due to trypsin inhibitors. Before using as food or animal feed, soybean seed must be heat-treated to break down trypsin inhibitor's activity.

Processed soybean foods include various products such as tofu (bean curd), yuba (soy protein extracts from soy milk), kinako (roasted soy powder), nimane (cooked beans), bean sprouts, and soymilk as non-fermented food products and soy sauce, miso, and natto as fermented food products. Among these soy-based food products, tofu, soy sauce, miso, bean sprouts and natto are predominant. Soybean oil has been traditionally appreciated as a high quality commodity in the world oil market and due to widely has used in many food and non-food applications.

Soybean oil is increasingly used in the food industry – for frying or baking foods, as a vegetable oil (table oil), as an ingredient in foods like salad dressings and margarins. The raw material is in the industry - soaps, creams, detergents, paints and varnishes, plastics, cleaners, medical preparations, ink. Lecitin from soybean oil has applications in the bakery, confectionary, pharmaceutical, textile and chemical industries. Soybean is also used to produce biodiesel. Biodisel *"methyl soyate"* is a renewable substitute for petroleum diesel with reduced greenhouse gas emissions. Results of numerous medical studies have indicated on the important role of soybean in the prevention and curing of chronic diseases. Healthy aspects of soyfoods go beyond the oil and protein and include minor compounds with nutraceutical properties.

Managing Foliar Diseases in Soybean

Integrated Pest Management

The use of foliar fungicides in soybean production has increased dramatically in the last 10 years. From 2004 to 2015, the percentage of planted soybean acres treated with a foliar fungicide rose from 1 percent to 11 percent. The cause of this increase is likely due to several factors, including increased disease prevalence, increased adoption of no-till, and recognized benefits of certain classes of fungicides to positively alter plant physiology. However, field trials still support using an integrated pest management (IPM) program to manage and control soybean diseases.

As part of an IPM program, it is best to consider foliar disease management prior to planting by selecting varieties that were disease resistant in previous years. Also, crop rotation should be a disease management strategy since inoculum can build up in crop residue under continuous soybean production. Despite these preventative strategies, foliar diseases can still reduce soybean yields, and foliar fungicides may be required to control a pathogen. Pathogens require specific modes of action and application timings for effective control, so it is important to understand how to recognize and differentiate soybean diseases that may be present in the field. The following is a description of four common foliar diseases that can reduce soybean yield in Nebraska and how to control them with fungicides and other management practices.

Disease Identification

Brown spot (*Septoria glycines*), Cercospora leaf blight (*Cercospora kikuchii*), frogeye leaf spot (*Cercospora sojina*), and white mold (*Sclerotinia sclerotiorum*) are foliar fungal diseases that should be monitored in Nebraska. Varieties vary in their response to these diseases, but each can significantly impact yield when conditions are favorable for disease development. Since fungicides can vary in their ability to control these diseases, it is important to properly identify which disease is present in the field. For example, bacterial pustule can easily be confused with soybean rust, and fungicides will not control bacterial pustule in soybeans. Making an incorrect identification could result in the wrong treatment for the disease.

Brown spot: Brown spot typically begins in the lower canopy during periods of high moisture or humidity. It is characterized by angular lesions with chlorotic margins (Fig. 18). Lesions can merge together to form large necrotic spots. Lesions will not have a pustule or visible spores. Yield losses typically range from 8 percent to 15 percent in severely affected fields. There are no known sources of genetic resistance, though some varieties may tolerate brown spot better than others. The best management strategy for brown spot is crop rotation.

Cercospora leaf blight: Cercospora leaf blight (CLB) affects leaves in the upper canopy (Figure 19). It favors hot temperatures and is characterized by a general bronzing or purpling of the upper leaf surface. Lesions are reddish purple and necrotic blotches form when the lesions coalesce. Yield loss can range from 15 percent to 30 percent nationally, but is typically less than 10 percent in Nebraska. The fungus causing CLB also causes purple seed stain, which can reduce seed quality and contribute to economic losses. Information on variety susceptibility to CLB is usually available from the seed company.

Frogeye leaf spot: Frogeye leaf spot typically develops on younger leaves in the upper canopy (Fig. 20). Lesions are circular with purple or reddish margins. Older lesions may develop dark centers where the fruiting bodies of the fungus are visible. Yield losses can be as much as 30 percent in the United States. However, losses are much lower in Nebraska—typically less than 10 percent. Crop rotation is an effective management practice, and resistant varieties are also available.

Sclerotinia Stem Rot (White Mold): Infection can occur when conditions are wet and humid, and temperatures are moderate during flowering. The fungus infects soybean flowers, and initial symptoms are visible during pod development. Early stem or pod water-soaked symptoms often initiate near colonized flowers, and leaves may wilt and turn gray-green before turning brown, curling, and dying. In a few days, diseased stem areas are killed and become tan and eventually bleached.

Infected plant parts generally have signs of the fungal pathogen as white, fluffy mycelium (fungal growth) during humid conditions and sclerotia on the surface of or embedded in the stem tissue (Figure 21). Yield can be reduced by 2 to 5 bu/ac for every 10 percent increase in incidence at the R7 growth stage. Tillage and short-term crop rotations will not reduce inoculum (pathogen) levels. Fungicide applications can help control the disease, but application timing during bloom is critical for good control. Other management practices that may help reduce disease incidence and severity are delayed planting, wide row spacing, reduced plant populations, and the use of lactofen-containing herbicides.

Resistance Management

Pesticide resistance to herbicides is widely discussed, but resistance to fungicides is also a growing problem. When a single mode of action is selected for disease control, or to enhance plant health in the case of quinone outside inhibitors (QoI), selection pressure is increased for development of resistant fungal strains. *Cercospora sojina*, the causal agent of frogeye leaf spot, is an example of a fungal pathogen that has been identified as having evolved resistance to QoI fungicides. The large amount of genetic diversity found in *C. sojina* allows the fungus to evolve fungicide resistance more rapidly when single modes of action products are used. As more active ingredients are becoming available as generic products, growers must realize the importance of using multiple modes of action even though some products are available at a lower cost. If we are not diligent in managing resistance, we could soon be faced with the same situation challenging us in weed control. Many herbicides are available, but we currently only have three fungicide modes of action that are commonly used commercially. Because the tools in our disease management toolbox for fungal diseases are more limited than those for weed management, it is imperative that we all be good stewards of fungicide use in our row crops.

Disease Control

Some foliar fungal diseases of soybean have similar life cycles, which means controlling one disease may help control other diseases. For example, the pathogens that cause brown spot, Cercospora leaf blight and frogeye leaf spot produce spores that overwinter in soybean residue. If crop rotation is used to manage frogeye leaf spot, incidence of brown spot will also be reduced the next time soybeans are grown in the field. Please refer to Table 1 for disease response to specific management practices.

Crop Rotation

Rotating to a nonhost crop after soybeans will interrupt the life cycle of certain fungi and reduce the amount of inoculum in the field for the next soybean crop. However, sclerotia of the white mold fungus can survive for a long time without a host crop, so rotations out of soybeans require several cropping seasons and the addition of cereal grain crops to reduce inoculum.

Tillage

Tillage after harvest will bury fungal-infested residue, which will break down more rapidly with soil contact. However, tillage practices may not be possible in no-till operations, or inadvisable if there is a potential for soil erosion.

Variety Selection

Many seed companies screen their genetic lines against specific diseases. Planting a variety resistant to a disease may greatly reduce potential yield losses. Ask your seed company representative for information concerning disease resistance in that company's varieties.

Fig. 18: Brown spot on soybeans showing lesions and yellowing leaves (A) and lower canopy defoliation (B).

Fig. 19. Cercospora blight on soybeans showing upper leaf surface (A) and upper canopy symptoms (B).

Fungicide Selection and Application

These diseases do not typically cause an economic yield loss except for white mold.

Fig. 20. Frogeye leaf spot on soybeans.

However, each can reduce yield when conditions for disease development are favorable. When market prices are higher, preventing small yield losses with a fungicide application may increase profitability. Field history, cropping sequence, variety susceptibility, and scouting should be used to determine the need for a fungicide. Two families of fungicides are primarily used for foliar disease control in soybeans in Nebraska. Quinone outside inhibitor (QoI) fungicides, formerly known as strobilurins, are most effective when they are applied before infection or at the very early stages of disease development.

Triazole fungicides can be applied after infection takes place but are most effective when applied at early stages of disease development. Succinate dehydrogenase inhibitors (SDHI) are a new mode of action that has come to market in limited products and will be important as we manage for fungicide resistance.

Many product labels recommend beginning fungicide application no earlier than the R1 (beginning flower) growth stage. In Nebraska, foliar diseases typically do not become a problem until the R3 (beginning pod) growth stage. Yield loss caused by foliar diseases primarily occurs from premature leaf drop, reducing the photosynthetic area and the amount of material the plant can produce to form grain, so fungicides must be applied at the correct time to prevent economic losses (Table 1).

Figure 21. Sclerotinia stem rot (white mold) on soybeans.

Table 1. Recommended control methods for foliar fungal diseases.

	Control Method				
	Crop Rotation	**Tillage**	**Variety**	**Fungicide1**	**Fungicide**
			Selection		Timing
Brown spot	YES	YES	YES2	YES	R3-early R4
Cercospora leaf blight	YES	YES	YES	YES	R3-R53
Frogeye leaf spot	YES	YES	YES	YES	R3-early R44
White mold	NO	NO	YES	YES	R1-R25

1. Follow fungicide label instructions for application rates and timings of product for effective control of each specific disease.

2. Varieties vary in susceptibility, but are not typically rated in variety information profiles.

3. Application timing for Cercospora leaf blight can be made through R5 with late disease onset, but the most probable returns can be expected from applications between R3 and early R4.

4. Application timing results vary and are not consistent for maximum returns with R3–R4 applications.

5. Best results expected with applications during bloom. Limited studies have shown benefits with applications at R3.

SUGAR BEET

Sugar beet, (*Beta vulgaris*), one of the four cultivated forms of the plant *Beta vulgaris* of the amaranth family (Amaranthaceae), cultivated as a source of sugar. Sugar beet juice contains high levels of sucrose and is second only to sugarcane as the major source of the world's sugar.

The sugar beet was grown as a garden vegetable and for fodder long before it was valued for its sugar content. Sugar was produced experimentally from beets in Germany in 1747 by chemist Andreas Marggraf, but the first beet sugar factory was built in 1802 in Silesia (now in Poland). Napoleon became interested in the process in 1811 because a British blockade had cut off the French Empire's raw cane sugar supply from the West Indies. Under his influence, 40 beet sugar factories were established in France. Although the industry temporarily collapsed after Napoleon's fall, it recovered in the 1840s. Beet sugar production then increased rapidly throughout Europe, and by 1880 the tonnage had overtaken that of cane sugar. Beet sugar now accounts for almost all sugar production in the European Union and for about one-fifth of total world production.

Historical Perspective

Sugar beet was introduced in India in 1950s. Exploratory trials for the root and seed crop were carried out by ICAR-Indian Institute of Sugarcane Research (IISR), Lucknow, at identified suitable sites in India. The first trial was conducted in India during 1959–1969 on exotic varieties of sugar beet with systematic multi-location testing with the suitable package and practices. This concept was established by All India Coordinated Research Project on Sugar beet by the Indian Council of Agricultural Research (ICAR) in 1970 which had three main centers at Pantnagar, Kanpur, and IISR Lucknow (Nodal Center).

ICAR established five sub-centers at Jalandhar, Hissar, Sriganganagar, Phaltan (Maharashtra) and Vegetable Research Station, Kalpa, and a new center at BCKV, Kalyani (WB) was added in the 7th five-year plan. In the view of research, 1970s and 1980s were the golden era of sugar beet with intense research activities in multiple aspects like agronomical practices, germplasm evaluation, varietal trial, entomological, pathological, and agricultural engineering.

The complete package and practices for sugar beet cultivation were developed along with several technologies of plant protection, cultural practices, and mechanization in sugar beet. The sugar beet cultivation practices in salt-affected soils were developed at Sundarbans (West Bengal) for the sub-tropical region of India, and found some varieties, viz., Pant S 1, Pant S 10, IISR Comp1-, IISR 2, and Magnapoly that were suitable to grow in alkaline soil conditions.

The utilization of sugar beet for alcohol production has been worked at Sundarbans. During 1976, All India Coordinated Research Project on sugar beet was transferred to G.B.P.U.A.&T, Pantnagar, but during VIII five-year plan, AICRP on Sugar beet was continued as Network Research Project with its headquarters at IISR, Lucknow, and five centers at Lucknow, Mukteshwar, Sriganganagar, Sundarbans (under the Kalyani Centre) and Kalpa.

Since 1970, sugar beet is being cultivated on a commercial scale at Sriganganagar in Rajasthan. A cane/beet sugar factory was established at Sriganganagar (Rajasthan) for commercial exploitation of sugar beet. The highest sugar recovery obtained from sugar beet was 11.32% during 1978–79 (Anonymous, 1978–79).

Development of diploid/polyploid varieties/hybrid was a major objective in sugar beet research at IISR, Lucknow and during 80's IISR developed two hybrids (LK HY-1 and LK HY-2), one composite (IISR Comp-1) and one synthetic (LKS-10). Multi-location testing of these hybrids was also conducted by the IISR, Lucknow.

During this era, breeder seed production of amelite variety LS-6 was done at Sugar beet Breeding Outpost, Mukteshwar. An open-pollinated diploid Russian variety seed namely Ramonskaya-06 (R-06) was found suitable to grow in India. Several exotic germplasm were evaluated against the best available check Ramonskaya (R-06). Some anisoploid varieties were also found suitable. At that time R-06 was under cultivation for several years but unfortunately, performance of this variety genetically deteriorated. So, IISR, Lucknow imported R-06 seed from USSR for genetically maintenance of this variety.

Under the strict supervision of breeders foundation seed and certified seed were produced at different centers. National Seeds Corporation (Sriganganagar) had undertaken the seed production of R-06 and after that Himachal Pradesh taken the production authority. Network Research Project on Sugar beet was merged with AICRP (Sugarcane) during December, 1994.

A joint meeting of AICRP (S) and Network Research Project on Sugar beet was held in 1998 at Vasantada Sugar Institute (VSI), Pune, and it was decided to phase out the sugar beet project because this crop was not much popularized as expected. In view of several benefits of sugar beet, IISR, Lucknow is still continuing the sugar beet research and seed production at its Outpost in Mukteshwar in Kumaon hills.

Development of tropicalized sugar beet varieties and seed production was taken up by multinational sugar beet seed companies under the direction of Syngenta Company in 2004. At the same time, ICAR also launched a Network Research Project to assess the feasibility of new sugar beet varieties under tropical conditions. Sporadic establishment of sugar beet processing units during 2008–2012 made the prospects of sugar beet look better up.

During this period a pilot plant was developed by VSI in Saamarth which later moved to Islampur. Ethanol plants were developed in Maharashtra and Karnataka and a sugar factory was established in Punjab.

Plant Disease Management

Powdery mildew first shows up on the older, lower leaves. The first symptoms are a wispy growth of white to light gray threadlike filaments, often radiating from a central point (Figures 22 and 23). Early symptoms of powdery mildew are not detected easily. Symptoms may be detected more readily under full sunlight, with the sun to your back. Once the early symptoms appear, the disease may progress very rapidly in favorable weather. A powdery white or gray-white growth may cover a leaf within a week, and the mildew may begin to appear on the middle leaves. Mildew may even appear on the upper leaves that have not completely expanded when the disease is severe. A characteristic odor similar to that of a musty basement may be noted in fields with severe powdery mildew.

Powdery mildew is likely to be more severe where nitrogen levels are low. When inspecting a field for powdery mildew, look in the areas of the field that are beginning to turn light green or yellow green due to depletion of available nitrogen. Powdery mildew is not likely to be found in the rest of the field if it cannot be found in the yellowing areas.

Severely mildewed leaves may begin to turn yellow within a month of initial symptoms.

Observations made near harvest time in North Dakota indicate that mildewed leaves are killed by light freezes that do not kill healthy leaves.

The sexual stage of the fungus has been observed in Colorado, Idaho, Nebraska, Wyoming and North Dakota. Structures known as ascomata are present in the sexual stage of the powdery mildew fungus.

To the naked eye, ascomata will appear as small, dark round spots that are mixed with the threadlike filaments of the fungus (Figure 24). With a hand lens, immature ascomata are yellow to orange, while mature ascomata are dark brown to black. The ascomata are circular. (Fig. 25).

Management

In the western U.S, varieties are evaluated for resistance to powdery mildew, and resistant varieties are available. Because powdery mildew rarely occurs in North Dakota and Minnesota, varieties are not evaluated for resistance to this disease in this area.

When powdery mildew occurs in late July or early August and it is not controlled, significant yield reduction occurs under severe disease conditions (Fig. 26). Fungicides are available to manage powdery mildew.

Fungicides in the strobilurin chemistry class, such as Gem, Headline (Fig. 27) and Quadris, can be used to manage powdery mildew. Sulfur fungicides such as Microthiol Disperss and Micro Sulf also can be used for powdery mildew management. The triazoles fungicides such as Proline, Inspire and Eminent also provide effective control of powdery mildew.

SORGHUM

Sorghum [*Sorghum bicolor* (L.)] is a cereal crop grown on nearly 43 million ha worldwide in the tropics, subtropics, and semiarid regions. Sorghum is grown for food, feed, forage, fiber, sugar, fuel, and alcoholic and nonalcoholic beverages. In the United States, sorghum is used primarily as a feed by the poultry, beef, and pork industries and a majority of U.S. sorghum is grown in semiarid regions, including Kansas and Texas. In the mid-Atlantic region of the United States, specifically in Virginia and North Carolina, sorghum production has increased because of the demand for locally sourced animal feed that cannot be met by current corn production. Sorghum produces greater yield relative to corn in harsher climatic conditions and on marginal land; thus, sorghum can be grown in drought-prone areas in the region that are not suitable for corn production. Sorghum production in Virginia and North Carolina increased from approximately 8,000 ha in 2005 to approximately 24,000 ha in 2016.

Anthracnose [*Colletotrichum sublineola* P. Henn in Kabát and Bubk (formerly *C. graminicola* (Ces.) G. W. Wilson)] is an economically important disease of sorghum worldwide. The fungus can infect all aerial parts of the plant, including the leaves, leaf sheaths, stalks, panicles, and grains. Foliar anthracnose is considered the most common and economically devastating stage of the disease because it interferes with photosynthate accumulation by covering the leaf surface, thereby reducing yield. Environmental conditions during the growing season in the mid-Atlantic are typically favorable for anthracnose development.

Disease outbreaks occur during periods of high rainfall or humidity >90% and when the temperature ranges from 25 to 30°C. Typical foliar symptoms appear as small, circular, elliptical, or elongated spots that have gray to straw-colored centers with visible setae from acervuli that can coalesce and cover the entire leaf at advanced stages. Anthracnose is a polycyclic disease and infection is enhanced by rain and contact of leaves with infested soil. In severe cases, defoliation and premature senescence occurs.

Under favorable conditions, development of anthracnose is more rapid from flowering through the grain filling stage compared with early developmental stages of sorghum. The pathogen can overwinter in crop debris as microsclerotia and mycelia, in infected seed, and on weed hosts such as johnsongrass [*S. halepense* (L.) Pers.].

Fig. 22. Sugar beet leaf showing symptoms of powdery mildew (left) vs. a healthy leaf (right)

Fig. 23. Close-up view of a sugar beet leaf infected by the powdery mildew fungus

A few prior studies have demonstrated sorghum yield losses to anthracnose and the potential for fungicides to reduce disease and protect grain yield. Anthracnose severity as low as 20% can impact yield and yield losses as high as 50% have been reported in susceptible sorghum cultivars. In sorghum producing regions in Nigeria and Brazil, use of fungicides (seed and foliar) has proven successful in managing this disease. Researchers reported that seed treated with fungicides, metalaxyl plus carboxin (Apron Plus 50DS) and thiram plus thiophanate methyl (Super Homai 70% WP), presented reductions in leaf anthracnose severity up to 70 days after planting compared with nontreated controls, but yield differences were not observed. Similarly, three foliar applications of benomyl at 7-day intervals beginning at disease onset resulted in lower anthracnose severity and increased yield compared with the nontreated control. also reported that three applications of the fungicides procloraz, carbendazim, benomyl, and azoxystrobin at 10-day intervals were effective in controlling anthracnose and increasing yield.

However, multiple foliar fungicide applications are unlikely to be economical for sorghum growers, and a single optimum application timing was not determined in these studies.

Diseases in semiarid regions are managed through breeding for genetic resistance and seed treatment fungicides, but foliar fungicides are not typically recommended because of the inconsistent occurrence and low severity of foliar diseases. However, with the subtropical conditions in the mid-Atlantic United States, anthracnose severity is greater than that in semiarid regions of the United States. A majority of the sorghum hybrids grown in the United States have been bred to be adapted to semiarid environments; thus, high-yield-producing hybrids may be susceptible to fungal diseases including anthracnose when grown in humid, wet environments.

Fig. 24. Sugar beet leaf with powdery mildew ascomata (sexual structures – dark spots)

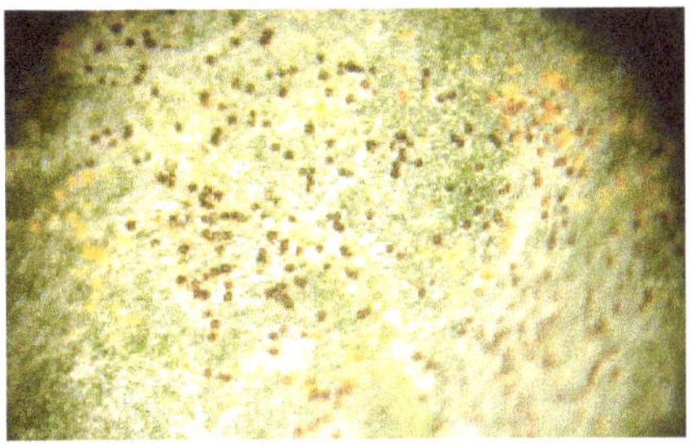

Fig. 25. Ascomata on sugar beet leaf as seen through a hand lens. Yellow and orange ascomata are immature; dark brown to black ascomata are mature

The effectiveness of foliar fungicides and optimum application timings have not yet been evaluated for sorghum anthracnose management in the mid-Atlantic region. Therefore, the objectives of this study were to (i) compare five fungicides and four application timings for anthracnose control and yield protection, (ii) estimate yield losses associated with sorghum anthracnose in the mid-Atlantic region, and (iii) evaluate the profitability of foliar fungicide applications for sorghum production in the region.

Fig. 26. Powdery mildew on plants not treated with a fungicide

Fig. 27. Powdery mildew-infected plants treated with Headline fungicide

Field trials

Field experiments were conducted over eight site-years in Virginia and North Carolina from 2015 to 2017. Anthracnose-susceptible sorghum hybrid 84P80 (DuPont Pioneer, Johnston, IA) was planted at a rate of approximately 193,000 seeds/ha. Information on previous crop, planting date, fungicide applications, date of anthracnose onset, and harvest date is provided in . Soil fertility, insects, and weeds were managed following standard practices for sorghum in the region. In 2016 and 2017, sugarcane aphids were controlled by using sulfoxaflor (as 0.1 liters of Transform/ha; Dow AgroSciences, Indianapolis, IN) or flupyradifurone (as 0.4 liters of Sivanto/ha; Bayer CropScience, Research Triangle, NC) and alternating the chemistry when multiple sprays were required.

Environmental data collection

Weather data for all North Carolina experimental sites were obtained from the North Carolina Climate Retrieval and Observations Network of the Southeast database.The environmental data for Rocky Mount and Lewiston-Woodville locations were recorded at on-site weather stations. Data for the Kinston location were obtained from a weather station at Cunningham Research Station located 7.1 km northeast from the experimental site. Weather data for the Suffolk site were obtained from Peanut-Cotton InfoNet, which reports data collected at the research farm where experiments were conducted.

Fungicides and application timings

The experiment included a 6 × 4 factorial arrangement of treatments in a randomized complete block design. The main factors were fungicide (nontreated control or one of five different fungicides) and application timing. Fungicides and rates were as follows: azoxystrobin (as 0.44 liters of Quadris 2.08 SC/ha; Syngenta Crop Protection, Greensboro, NC), pyraclostrobin (as 0.44 liters of Headline SC/ha; BASF Corporation, Research Triangle Park, NC), picoxystrobin (as 0.44 liters of Aproach SC/ha; DuPont Crop Protection, Wilmington, DE), pyraclostrobin plus fluxapyroxad (as 0.29 liters of

Priaxor 4.17 SC/ha; BASF Corporation), and propiconazole (as 0.29 liters of Tilt 3.6 EC/ha; Syngenta Crop Protection). The four application timings were boot (head enclosed in a swollen leaf sheath), flower (beginning of flowering in 50% of the plants), late (14 days after flower application), or applications at all three stages. Each site had three or four replications with 2.4-m fallow alleys between blocks. Each treatment plot consisted of four rows that were 9 m long by 3.7 m wide except for the 2016 and 2017 experiments in Kinston, North Carolina (KS-16 and KS-17, respectively), which had three rows with dimensions of 9 m by 2.8 m. Anthracnose severity was evaluated biweekly as a visual estimate of the percentage of leaf area with symptoms and signs of sorghum anthracnose starting at the boot stage.

At the sites in Virginia, fungicide treatments were applied with a Lee Spider Sprayer having eight 8002VS nozzles spaced 46 cm apart delivering 186 liters/ha at 262 kPa. At the sites in North Carolina, fungicide treatments were applied with a backpack sprayer having eight XR 11002 flat fan nozzles spaced 46 cm apart delivering 140.3 liters/ha at 207 kPa. At the end of the season, grain was harvested from the center two rows of each plot for all of the sites except for Kinston, where all three rows of a plot were harvested, using a mechanical harvester equipped with a grain gauge and moisture meter. Yield, moisture, and test weight were measured from each plot at harvest, and yield/hectare was calculated based on a standard moisture content adjusted to 14%.

Profitability calculations

The profitability of fungicide applications in sorghum was estimated in U.S. dollars (USD) based on average local prices from three retailers each in Virginia and North Carolina for Priaxor ($120.2/liter), Headline ($108.6/liter), Quadris ($48.2/liter), Aproach ($67.0/liwter), and Tilt ($22.1/liter). The estimated machinery and labor costs for fungicide applications were $49.3/ha and $33.8/ha, respectively, and the projected price of grain sorghum in Virginia and North Carolina for the year 2018 was $0.17/kg. The increase in yield required to cover the cost of use of a fungicide (break-even yield) was determined using the following formula, modified from:

Yield (in kg/ha)required to cover the cost of using a fungicide = [Fungicide cost (in USD) + Machinery and labor cost for fungicide spray(in USD) per hectare]/[Grain sorghum price (in USD/kg)]

Statistical analyses

In some experiments, disease onset was late in the season so disease severity could not be evaluated until the hard dough stage, when grains changed color from green to brown and were hard when squeezed. Hence, only the final disease severity evaluation at the hard dough stage was used for statistical analysis and comparison. An initial analysis was performed using block, fungicide, application timing, site-year, and interactions among these factors for anthracnose severity in the model. The effect of site-year and its interactions with fungicide and application timing were significant, so each experiment was analyzed separately. Percentage anthracnose severity data were arcsine square root transformed prior to analysis to meet the assumptions of analysis of variance.

Data for anthracnose severity and yield were analyzed using the general linear model procedure (glm) in R software version 3.2.2 using the package, agricolae. In addition, regression analyses of environmental variables and disease and anthracnose severity and yield were conducted using the linear model procedure (lm) in agricolae. Anthracnose severity was subjected to ANOVA using a model that included block, fungicide, application timing, and their interactions as sources of variation. Means were compared using the Tukey honest significant difference test at the significance level of á = 0.05, and back-transformed values are presented.

Results

Date of foliar anthracnose onset and rate of disease development in the nontreated controls varied among experiments. Site-years where disease onset was prior to the boot stage had the most rapid rate of disease development between planting and flowering. However, there was a poor relationship between anthracnose severity at the hard dough stage and either disease onset ($P = 0.15$, $r = 0.56$-) or disease development prior to flowering ($P = 0.40$, $r = 0.35$). Percent yield losses attributable to disease at each site-year were estimated based on the difference between the nontreated control and the greatest yield-producing fungicide treatment.

Among the disease variables, disease onset relative to planting was most predictive of yield losses (P = 0.047, r = -0.71) with earlier disease onset resulting in greater yield losses.

Relationships between environmental variables, disease, and yield losses were examined for the eight experiments. Mean monthly temperatures were similar among site-years; however, total precipitation differed, with some sites receiving more precipitation than others during the growing season. There was no relationship between anthracnose severity at the hard dough stage and monthly rainfall, but greater rainfall the month prior to flowering was associated with earlier disease onset ($P = 0.068$, $r = 0.67$-), more rapid rates of disease development prior to flowering ($P = 0.007$, $r = 0.85$), and greater percent yield losses ($P = 0.034$, $r = 0.74$). Notably, even though the Rocky Mount site in 2016 (RM16-) only had moderate anthracnose severity at the hard dough stage, it had 2.5 to 11 times greater rainfall compared with other site-years the month prior to flowering, as well as the earliest disease onset, the greatest rate of disease development prior to flowering, and the greatest percent yield losses among the eight experiments.

Table 2. Monthly rainfall and average temperatures during the sorghum growing season for eight site-years in Virginia and North Carolina

Site[z]	May	June	July	August	September	October
Rainfall (mm)						
SF-15	12	161	96	55	115	74
RM-15	60	84	54	82	167	137
RM-16	94	111	491	85	246	259
KS-16	175	112	132	128	240	84
SF-17	120	89	60	186	94	88
RM-17	125	122	151	176	76	87
LW-17	148	143	189	128	82	69
KS-17	93	120	43	86	108	77
Mean temperature (°C)						
SF-15	21	26	26	25	23	16
RM-15	21	26	26	25	23	16
RM-16	20	24	27	27	23	18
KS-16	20	25	27	27	24	19
SF-17	20	26	28	25	22	21
RM-17	21	24	26	25	22	18
LW-17	20	24	26	24	22	18
KS-17	22	25	27	25	22	18

[z]Experiment locations were as follows: SF = Suffolk, Virginia (Tidewater Agricultural Research and Extension Center); RM = Rocky Mount, North Carolina (Upper Coastal Plain Research Station); KS = Kinston, North Carolina (Casewell Development Center Research Farm); and LW = Lewiston-Woodville, North Carolina (Peanut Belt Research Station). Numbers 15, 16, and 17 following the location abbreviation indicate the years 2015, 2016, and 2017, respectively.

Anthracnose severity varied among fungicide treatments at all site-years, but the impact of application timing and the interaction between fungicide and timing varied among experiments. Overall, pyraclostrobin and pyraclostrobin plus fluxapyroxad provided the greatest reduction in anthracnose severity, followed by azoxystrobin; picoxystrobin and propiconazole treatments were not significantly different from the nontreated control regardless of site-year. When application timing had an impact on disease severity, three fungicide applications provided the greatest level of disease control, and fungicides applied 2 weeks after flowering resulted in the least control. In two experiments, there was a significant fungicide × application timing interaction. Pyraclostrobin and pyraclostrobin plus fluxapyroxad reduced disease severity when applied at the boot, flower, or all three timings, but fungicides did not reduce anthracnose severity when applied 2 weeks after flowering.

Table 3.Impact of fungicides and application timings on foliar anthracnose severity in sorghum hybrid P84P80 at eight site-years in Virginia and North Carolina

Variable[y]	Anthracnose severity (%)[z]							
	SF-15	RM-15	RM-16	KS-16	SF-17	RM-17	LW-17	KS-17
Fungicide (F)								
NTC	71.9 a	63.1 a	44.7 a	16.7 a	43.4 a	27.4 a	82.5 a	4.7 a
PROP	71.2 a	61.8 a	42.8 ab	12.9 a	42.2 a	24.2 ab	79.4 ab	8.4 a
AZOX	72.6 a	62.4 a	38.7 b	9.9 ab	27.3 b	14.7 b	64.4 b	3.5 ab
PICO	72.6 a	65.9 a	42.1 ab	10.5 a	37.0 ab	18.3 ab	77.8 ab	3.7 ab
PYRA	70.6 a	62.4 a	25.9 c	1.5 c	2.3 d	2.3 c	23.7 c	0.5 b
PYRA + FLUX	63.8 b	52.8 b	22.4 c	2.5 bc	8.6 c	2.8 c	28.3 c	0.5 b
P value	<0.01	<0.01	<0.01	<0.01	<0.01	<0.01	<0.01	<0.01
Application (A)								
Boot	70.0	63.1	39.0 ab	10.1	25.6 ab	16.7 a	56.0 b	3.6
Flower	71.7	59.9	36.5 b	9.4	24.2 ab	14.5 ab	61.6 b	2.7
Late	71.7	60.8	40.6 a	8.6	29.7 a	13.3 ab	78.7 a	4.2
All	68.6	61.8	27.5 c	4.5	17.5 b	8.5 b	42.2 c	1.6
P value	0.21	0.52	<0.01	0.09	<0.01	0.02	<0.01	0.18
F × A								
P value	**0.42**	**0.35**	**<0.01**	**0.24**	**0.6**	**0.08**	**<0.01**	**0.57**

[y]Tested variables included fungicides (F) and application timings (A). Significant treatment effects and treatment interactions (F × A) were determined using a factorial analysis of variance. NTC = nontreated control; PROP = propiconazole (as Tilt); AZOX = azoxystrobin (as Quadris); PICO = picoxystrobin (as Aproach); PYRA = pyraclostrobin (as Headline); and PYRA + FLUX = pyraclostrobin + fluxapyroxad (as Priaxor).

[z]Anthracnose severity was rated as a visual estimate of the percent leaf area with signs and symptoms of sorghum anthracnose. Experiment locations were as follows: SF = Suffolk, Virginia (Tidewater Agricultural Research and Extension Center); RM = Rocky Mount, North Carolina (Upper Coastal Plain Research Station); KS = Kinston, North Carolina (Casewell Development Center Research Farm); and LW = Lewiston-Woodville, North Carolina (Peanut Belt Research Station). Numbers 15, 16, and 17 following the location abbreviation indicate the years 2015, 2016, and 2017, respectively. Within a treatment factor, means followed by the same letter are not significantly different from each other according to the Tukey honest significant difference test.

Effects of fungicides and application timings on yield varied by experiment. Differences in yield between the nontreated control and fungicide treatments were detected at only three of the eight site-years, and only the pyraclostrobin and pyraclostrobin plus fluxapyroxad treatments resulted in greater yield. Three applications resulted in numerically greater yield compared with single applications, but overall, boot and flower applications had greater yield compared with late fungicide applications. For RM-16, the experiment with the earliest disease onset and greatest yield losses, there was a significant fungicide × application timing interaction and pyraclostrobin and pyraclostrobin plus fluxapyroxad protected yield when applied at flowering or all three timings but did not protect yield when applied at boot or 2 weeks after flowering.

For three experiments, there was not a relationship between leaf anthracnose severity and yield. For the Kinston experiments in 2016 and 2017, there was a significant but poor correlation between disease severity and grain yield. There was a negative linear correlation of leaf anthracnose severity and yield for three experiments, with correlation coefficient (r) values ranging from +0.75 to +0.86 and yield losses ranging from 27 to 85 kg/ha attributable to each percentage of leaf area with anthracnose.

Two fungicides, pyraclostrobin and pyraclostrobin plus fluxapyroxad, protected yield compared with the nontreated control and other fungicides tested in this study. Hence, the profitability calculations were performed for a single application of these two fungicides. Because the late application of fungicides did not reduce anthracnose severity or protect yield compared with the nontreated control, the profitability calculations were conducted for boot and flower applications only.

A single application of pyraclostrobin resulted in yield 49 to 2,448 kg/ha greater than the nontreated control, and a single application of pyraclostrobin plus fluxapyroxad resulted in yields 111 to 2,694 kg/ha greater than the nontreated control. To cover the cost of a single fungicide application, a yield increase of 769 and 693 kg/ha was needed for pyraclostrobin ($130.70/ha) and pyraclostrobin plus fluxapyroxad ($117.80/ha), respectively.

Even though the only significant yield difference between fungicide applied at the boot stage and the nontreated control was for pyraclostrobin plus fluxapyroxad at a single site-year, the value of the difference in yield between a single application at boot compared with the nontreated control exceeded the cost of fungicide and resulted in a positive return on investment ranging from $13 to $307/ha in four of the experiments.

When these fungicides were applied at the flowering stage, both fungicides had greater yield compared with the nontreated control at two site-years, but there was a positive net return ranging from $10 to $373/ha from a single fungicide application for six of the eight experiments.

Time of disease onset and environmental conditions varied among experiments, and this likely impacted the levels of anthracnose control and yield response to fungicides observed in this study. Among the eight experiments, RM-16 received the greatest amount of rainfall prior to flowering, had the earliest disease onset relative to planting of the sorghum crop, and had the greatest anthracnose severity between boot and flowering.

In addition, RM-16 was the site-year with the greatest yield losses in the absence of anthracnose control with fungicides. This suggests that early rainfall prior to the boot stage of the sorghum crop is a risk factor for anthracnose disease development and yield losses. Dissemination and germination of *C. sublineola* is water dependent and expansion of the anthracnose lesion and formation of acervuli is favored by temperatures near 30°C. Average monthly temperature of 25°C or greater and high precipitation events likely contributed to early disease onset and progression and yield losses at some locations in the current study. For experiments in Suffolk, Virginia, and Lewiston, North Carolina, in 2017 (SF-17 and LW-17, respectively), disease onset was early, but the progression was slow until flowering. As the crop entered reproductive stages, disease severity increased rapidly, which is similar to previous reports where slow anthracnose development was observed prior to flowering.

Despite the slower disease development for the SF-17 and LW-17 experiments compared with RM-16, all three site-years had yield losses in the absence of disease control with fungicides. This suggests that early disease onset, which is associated with early-season rain events, may be more predictive of potential yield losses and need for fungicides than severity of anthracnose observed at flowering of the sorghum crop. Hence, regular scouting is suggested at early developmental stages when plants are close to boot stage, and decisions to apply a fungicide should be based on the presence of disease. Results from this study indicate that fungicide application at the flowering stage is critical for disease control and to protect yield.

Application of fungicides may not be profitable if disease is absent prior to the flowering stage, as we did not see any significant yield differences between nontreated and treated plots under low disease severity or when disease onset occurred after flowering. Our results are in accordance with two other studies that showed that application of fungicides at low or no disease conditions does not increase grain yields in agronomic crops.

Yield protection with fungicide treatments was only observed for three experiments that had early disease onset followed by relatively high levels of foliar anthracnose in nontreated controls. Even though boot and flowering fungicide applications were applied 3 to 18 days after disease onset, they effectively protected reduced anthracnose severity and protected yield. This is likely because initial infections were observed on the lower leaves and did not reach the flag leaf until several weeks after disease onset; thus, fungicide applications between the boot and flowering stages protected the upper canopy that provides a majority of the photosynthates for the developing grain. For the SF-15 and RM-15 experiments, anthracnose severity was high, but yield did not differ among fungicides or application timings, which could be a result of heavy precipitation toward the end of the growing season that resulted in lodging, shattering of grains, and sprouting in heads (*personal observation*).

For the KS-16 and RM-17 experiments, there was no difference in yield among fungicide treatments, which could be attributable to lower disease severity at these site-years. The Pioneer 84P80 hybrid used in this study is highly susceptible to anthracnose but also has high yield potential. Hence, under low disease pressure, this hybrid generally produced high grain yield. This indicates that application of fungicides in the absence of disease or when disease pressure is low does not produce a yield benefit.

Leaf anthracnose severity as low as 20% is reported to impact sorghum grain yield. Our study also shows a significant negative correlation between leaf anthracnose severity and yield. Based on our results, for every percent increase in leaf anthracnose, there is a 27 to 85 kg/ha loss in grain yield, which is a greater range compared with a recent study from Brazil that reported a 11 to 39 kg/ha reduction in yield for each percent increase in anthracnose severity in susceptible sorghum genotypes.

The greater range in yield reduction might be attributable to the greater number of experimental sites in this study. This illustrates the importance of managing sorghum anthracnose in the mid-Atlantic and other regions where environmental conditions favor fungal disease development.

Pyraclostrobin and pyraclostrobin plus fluxapyroxad were the only two fungicides that consistently reduced anthracnose severity, protected yield, and increased farm profitability with a single application at flowering. Hence, inclusion of a single application of a pyraclostrobin-based fungicide is likely to be beneficial when anthracnose onset is early and environmental conditions are favorable for disease development. Application of these fungicides after flowering did not protect yield, and applications following this stage of crop development are unlikely to be profitable to growers.

The use of pyraclostrobin plus fluxapyroxad is more economical than pyraclostrobin even though the price of pyraclostrobin plus fluxapyroxad per liter ($120.2/liter) is greater compared with that of pyraclostrobin ($111.0/liter). This is because a smaller quantity of pyraclostrobin plus fluxapyroxad (0.29 liters/ha) is enough to reduce anthracnose severity, protect yield, and increase farm income compared with 0.44 liters of pyraclostrobin/ha.

Even though pyraclostrobin-containing fungicides provided the most consistent anthracnose control and yield protection, a single application or three applications of azoxystrobin reduced anthracnose severity in some experiments. This is in contrast to a study in Brazil that identified three applications of azoxystrobin as one of the most effective fungicides to control sorghum anthracnose. However, single applications of fungicides were not evaluated in this study.

Regardless, based on this and the current study, azoxystrobin may be an alternative if pyraclostrobin and pyraclostrobin plus fluxapyroxad are not available, whereas propiconazole and picoxystrobin did not reduce anthracnose severity under field conditions, even after three applications. The limited number of effective fungicide active ingredients and modes of action labeled for control of sorghum anthracnose indicates a high risk for fungicide resistance development within populations of *C. sublineola*.

The fungicide active ingredient pyraclostrobin is in the quinone outside inhibitor class, which is categorized as a high risk for fungicide resistance development especially if it is not rotated with different fungicide modes of action. Fungicides, premixes, or tank mixes with different modes of action to rotate with pyraclostrobin should be explored to reduce the potential development of fungicide resistance. The fungicide Priaxor is a mixture of pyraclostrobin and fluxapyroxad. Fluxapyroxad is a member of the succinate dehydrogenase inhibitor class of fungicides.

This premix fungicide may have a relatively reduced risk of resistance development compared with pyraclostrobin alone, but it will still be important in the future for additional fungicide modes of action to be labeled for anthracnose control in sorghum.

The climate of the mid-Atlantic is warm and humid compared with the major sorghum growing regions in the United States and provides suitable conditions for anthracnose epidemics that may severely impact yield and farm profitability. Anthracnose development and dispersal is temperature and precipitation dependent. Based on the current study conducted in the mid-Atlantic United States, when average temperatures e+25°C are accompanied by frequent or heavy rainfall events, anthracnose onset occurs earlier and more rapid rates of disease development are observed.

If anthracnose symptoms and signs are observed at or before the boot stage, there is a greater chance of disease outbreaks and yield losses later in the season. Thus, the sorghum crop should be scouted for disease when the crop is approaching the boot stage. A single application of fungicide (pyraclostrobin or pyraclostrobin plus fluxapyroxad) is warranted if anthracnose onset is at or prior to the boot stage and should be applied no later than flowering to reduce anthracnose severity, protect yield, and increase farm profitability, particularly when a susceptible hybrid is being grown. In the absence of anthracnose prior to flowering, fungicide application is unlikely to protect yield or farm profitability. In this study, we did not consider the effect of host resistance on management of sorghum anthracnose.

SUNFLOWER

Sunflower, (genus *Helianthus*), genus of nearly 70 species of herbaceous plants of the aster family (Asteraceae). Sunflowers are native primarily to North and South America, and some species are cultivated as ornamentals for their spectacular size and flower heads and for their edible seeds. The Jerusalem artichoke (*Helianthus tuberosus*) is cultivated for its edible underground tubers.

Sunflowers are botanically classified as *Helianthus annuus* L. They are large plant and are grown throughout the world because of their relatively short growing season. Sunflower is an annual herb, with a rough, hairy stem, 3 to 12 feet high, broad, coarsely toothed, rough leaves, 3 to 12 inches long and circular heads of flowers, 3 to 6 inches wide in wild specimens and often a foot or more in cultivation. The flower-heads are composed of many small tubular flowers arranged compactly on a flattish disk: those in the outer row have long strap-shaped corollas, forming the rays of the composite flower.

Each sunflower head, or inflorescence, is actually composed of two types of flowers. What appears to be yellow petals around the edge of the head are actually individual ray flowers. The face of the head is comprised of hundreds of disk flowers, which each form into a seed (achene).

The basic chromosome number for the Helianthus genus is 17. Diploid, tetraploid and hexaploid species are known. There are only 14 annual species of Helianthus. Plant breeders have made interspecific crosses within the genus and have transferred such useful characters as higher oil percentage, cytoplasmic male sterility for use in production of hybrids, and disease and insect resistance to commercial sunflower.

History

Sunflower (*Helianthus annuus* L.) is one of the few crop species that originated in North America (most originated in the fertile crescent, Asia or South or Central America). It was probably a "camp follower" of several of the western native American tribes who domesticated the crop (possibly 1000 BC) and then carried it eastward and southward of North America. The first Europeans observed sunflower cultivated in many places from southern Canada to Mexico.

Sunflower was probably first introduced to Europe through Spain, and spread through Europe as a curiosity until it reached Russia where it was readily adapted. Selection for high oil in Russia began in 1860 and was largely responsible for increasing oil content from 28% to almost 50%. The high-oil lines from Russia were reintroduced into the U.S. after World War II, which rekindled interest in the crop.

However, it was the discovery of the male-sterile and restorer gene system that made hybrids feasible and increased commercial interest in the crop. Production of sunflowers subsequently rose dramatically in the Great Plains states as marketers found new niches for the seeds as an oil crop, a birdseed crop, and as a human snack food. Production in these regions in the 1980s has declined mostly because of low prices, but also due to disease, insect and bird problems. Sunflower acreage is now moving westward into dryer regions; however, 85% of the North American sunflower seed is still produced in North and South Dakota and Minnesota.

Production

In recent years, the sunflower cultivated area has been steadily increasing due to the breeding of dwarf high yielding hybrids that also facilitate mechanization and the emphasis given to polyunsaturated acids for human consumption. Global production grew steadily in last 25 years (PSD-USDA, 2011), and FAO expect a total world output close to 60 million tons towards 2050. The four largest producers (Russia, Ukraine, European Union and Argentina) account for 70% of global volume, with an exponential growth of production in the last ten years in the Black Sea region, with increased acreage an higher yields achieved by the replacing old varieties by hybrid seeds.

Uses:

Edible oil

Commercially available sunflower varieties contain from 39 to 49% oil in the seed. In 1985-86, sunflower seed was the third largest source of vegetable oil worldwide, following soybean and palm. The growth of sunflower as an oilseed crop has rivaled that of soybean, with both increasing production over 6-fold since the 1930s. Sunflower accounts for about 14% of the world production of seed oils (6.9 million metric tons in 1985-86) and about 7% of the oilcake and meal produced from oilseeds. Europe and the USSR produce over 60% of the world's sunflowers.

The oil accounts for 80% of the value of the sunflower crop, as contrasted with soybean which derives most of its value from the meal. Sunflower oil is generally considered a premium oil because of its light color, high level of unsaturated fatty acids and lack of linolenic acid, bland flavor and high smoke points. The primary fatty acids in the oil are oleic and linoleic (typically 90% unsaturated fatty acids), with the remainder consisting of palmitic and stearic saturated fatty acids. The primary use is as a salad and cooking oil or in margarine. In the USA, sunflower oils account for 8% or less of these markets, but in many sunflower-producing countries, sunflower is the preferred and the most commonly used oil.

High oleic sunflower oil (over 80% oleic acid) was developed commercially in 1985 and has higher oxidated stability than conventional oil. It has expanded the application of sunflower oils for frying purposes, tends to enhance shelf life of snacks, and could be used as an ingredient of infant formulas requiring stability.

Meal

Non-dehulled or partly dehulled sunflower meal has been substituted successfully for soybean meal in isonitrogenous (equal protein) diets for ruminant animals, as well as for swine and poultry feeding. Sunflower meal is higher in fiber, has a lower energy value and is lower in lysine but higher in methionine than soybean meal. Protein percentage of sunflower meal ranges from 28% for non-dehulled seeds to 42% for completely dehulled seeds. The color of the meal ranges from grey to black, depending upon extraction processes and degree of dehulling.

Industrial Applications

The price of sunflower oil usually prohibits its widespread use in industry, but there are several applications that have been explored. It has been used in certain paints, varnishes and plastics because of good semidrying properties without color modification associated with oils high in linolenic acid. In Eastern Europe and the USSR where sunflower oil is plentiful, sunflower oil is used commonly in the manufacture of soaps and detergents. The use of sunflower oil (and other vegetable oils) as a pesticide carrier, and in the production of agrichemicals, surfactants, adhesives, plastics, fabric softeners, lubricants and coatings has been explored. The utility of these applications is usually contingent upon petrochemical feedstock prices.

Sunflower oil contains 93% of the energy of US Number 2 diesel fuel (octane rating of 37), and considerable work has been done to explore the potential of sunflower as an alternate fuel source in diesel engines. Blends of sunflower oil and diesel fuel are expected to have greater potential than the burning of pure vegetable oil.

Non-Oilseed

The use of sunflower seed for birdfeed or in human diets as a snack, has grown consistently over the past 15 years. Varieties used for non-oilseed purposes are characterized by a larger seed size and require slightly different management practices. During processing, seed is divided into 1) larger seed for in-shell roasting, 2) medium for dehulling, and 3) small for birdseed. Standards for different uses vary.

Forage

Sunflower can also be used as a silage crop. It can be used as a double crop after early harvested small grains or vegetables, an emergency crop, or in areas with a season too short to produce mature corn for silage.

Forage yields of sunflower are generally less than corn when a full growing season is available. In one study, sunflower dry matter yields ranged from 2.0 to 3.0 ton/acre compared with 3.1 to 3.8 ton/acre for corn. Moisture content of sunflower at maturity is usually high (80 to 90%) and would require wilting before ensiling.

Effect of Fungicide and Timing of Application

Sunflower rust, caused by *Puccinia helianthi* Schwein., is an important yield-limiting disease in United States sunflower production areas. The pathogen is macrocyclic, autoecious, and can overwinter in all areas of sunflower production in the United States. Consequently, the pathogen has potential for frequent sexual reproduction, which can result in race changes. Similarly, earlier onset of rust in producers' fields can lead to the development of greater year-end severity levels and yield loss. Sunflower rust incidence has increased in North Dakota since 2008 and yield losses of up to 80% have been documented. Most commercially grown hybrids are susceptible to common races of rust, and the impact of crop rotation on rust is limited because of long-distance spore dispersal and ubiquitous wild sunflower populations harboring the pathogen, leaving fungicides as one of the only effective management tools.

The majority of the studies evaluating the performance of fungicides for sunflower rust management were conducted outside the United States. Results from efficacy trials conducted in Kenya and Turkey demonstrated that fungicides increased yield-component values when compared with nontreated controls.

In the 1990s in Israel, researchers developed a fungicide action threshold based on both leaf severity and growth stage and recommended a fungicide application when leaf severity of 3% on the upper four leaves is reached prior to 27 days after flowering. However, development of the threshold was based solely on research conducted with demethylation inhibitors, and in a production system different from the north-central United States (humidity, day length, and temperature). Since then, quinone outside inhibitors have been labeled, and are recommended for management of other rust pathosystems, but efficacy and timing for sunflower rust has not been investigated. The objectives of this study were to evaluate fungicide efficacy and timing for management of sunflower rust and protection of yield.

Locations and experimental design

Experiment sites were at four locations in North Dakota: Casselton (CASS), the NDSU Carrington Research Extension Center (CREC), Grandin (GRAN), and the NDSU Langdon Research Extension Center (LREC). Trials were conducted in a randomized complete block design with four replications at CASS in 2008, CREC from 2008 to 2011, GRAN from 2009 to 2011, and LREC from 2008 to 2011. CREC was under center-pivot irrigation, all other locations were nonirrigated. Four row plots were seeded with a susceptible confection hybrid 'Jaguar' (Seeds2000, Breckenridge, MN). Row width was 76.2 cm and seed spacing was 21.6 cm at each location. Four row plots were used at all locations; the middle two rows of each plot were used for data collection, while the other plot rows were used as borders to interplot interference. Planting date, rust detection date, and plot length varied at each location. Agronomic practices appropriate for sunflower production were followed for each location and year.

Artificial inoculation.

In order to facilitate disease development, plots were artificially inoculated at each location. Inoculum was produced by increasing urediniospores of *P. helianthi* isolate ND01-07 (race 336) on a greenhouse-grown susceptible sunflower hybrid to ensure a fresh viable source of inoculum. Collected spores were suspended in Soltrol 170 and quantified to approximately 275,000 spores/ml. The spore suspension was applied using a backpack mounted leaf blower (model # BR600, Stihl Incorporated USA, Virginia Beach, VA) with small modifications to the air wand. A 0.5 cm diameter hole was drilled approximately 13.7 cm from the mouth of the wand, on the top side. A 0.5 cm outside-diameter hose was inserted approximately 1 cm into the hole in the wand and secured with adhesive tape (item # 283227, ShurTech Brands, Avon, OH). The urediniospore suspension was injected through the hose and into the airstream where it vaporized and was dispersed onto plant tissue. In 2008 and 2009, only border rows of the plots and internal spreaders plots were inoculated, while in 2010 and 2011, the treatment rows were inoculated. Timing of inoculation coincided with disease onset objectives. Inoculation was completed during early sunflower reproductive stages in 2008 and 2009; and in late vegetative stages during 2010 and 2011.

Fungicide efficacy

Efficacy trials were established in 2008 and 2009. Locations varied in the total number of fungicides evaluated but all trials included the demethylation inhibitors (DMI - FRAC 3); prothioconazole (Proline, Bayer CropScience, Research Triangle Park, NC), prothioconazole + tebuconazole (Prosaro, Bayer CropScience), tebuconazole (Tebuzol, United Phosphorous Inc., King of Prussia, PA), and the quinone outside inhibitors (QoI - FRAC 11); pyraclostrobin (Headline, BASF, Research Triangle Park, NC) and azoxystrobin (Quadris, Syngenta Crop Protection, Greensboro, NC).

Other DMI fungicides included at some locations were metconazole (Quash, Valent USA, Walnut Creek, CA) and tebuconazole (Folicur, Bayer CropScience) (Table 1). Succinate dehydrogenase inhibitors (SDHI - FRAC 7) included at some locations were boscalid (Endura, BASF) and penthiopyrad (Vertisan, DuPont Agricultural Products, Wilmington, DE). Additionally, a nontreated control (NTC) was used at each location. All fungicides were applied between growth stages R5.0 (start of flowering) and R5.5 (midflowering).

Fungicide timing—Normal onset.

Timing trials conducted in 2008 and 2009 were designed to simulate the onset of rust in a "normal" disease onset scenario. We define a normal disease onset as when rust is first found on the upper leaves of sunflowers during the reproductive stages of development. This approximately simulates the time of disease onset in production fields when the source of inoculum are urediniospores from nearby fields or long distance dispersal. To simulate this, spreader rows were inoculated with urediniospores approximately 1 to 2 weeks before the reproductive growth stages began in the trials. Fungicide applications were made at sunflower growth stages R3-4 (midbud elongation), R5 (flowering), and R6 (flowering complete) using single or multiple applications of pyraclostrobin or tebuconazole. An NTC was used at each location as well. The number of treatments varied among location and years.

Fungicide timing—Early onset.

Timing trials conducted in 2010 and 2011 were designed to simulate an "early" disease onset scenario. We define early disease onset as when rust is found on the upper portion of the plant prior to growth stage R1. This approximately simulates the time of disease onset in production fields when the source of inoculum originates from within the field. To simulate this, all plots were initially inoculated with urediniospores between V6 and V12 growth stages, and reinoculated periodically until uredinia were visually observed.

Fungicide applications were made at three growth stages: V8-V12 (late vegetative), R1 (bud formation), and R5 (bloom) using single and multiple applications. Experimental timings were developed, in part, to coincide with sunflower production practices. Herbicide applications are usually initiated at late vegetative (V8-V12), fungicide applications for nondisease yield impacts have been suggested to be applied at R1, and the R5 timing was deemed the most important fungicide timing of the three tested in 2008 and 2009. Pyraclostrobin was used at all locations, while tebuconazole was also evaluated at GRAN.

Fungicide application

Fungicides were applied at 241.3 kPa with CO_2 powered backpack sprayers at GRAN, CASS, and LREC and with a tractor sprayer at CREC. Applications were made at 187 liters/ha at GRAN and CASS, 86 liters/ha at LREC, and 121.6 liters/ha at CREC. Teejet 8002 flat fan nozzles were used at all locations. Nozzle spacing was 76.2 cm at CASS, GRAN, and LREC and 38.1 cm at CREC. The target area of application was the upper four leaves of the sunflower plant.

Data collection

Disease evaluations were conducted visually, with the aid of rust assessment diagrams. The sunflower rust assessment diagrams are more conservative than other rust assessment tools. For example, a 1% rust severity on sunflower rust diagrams translates to approximately a 20% severity for wheat rust using the modified Cobb scale (Peterson et al. 1948). The mean percent rust severity on the upper four leaves of 10 randomly selected plants in the center two rows of each plot were evaluated (Shtienberg 1995). Assessments were conducted at approximately 2 week intervals beginning at disease onset and ending at plant maturity (R8-R9). Disease severity ratings were used to calculate the area under disease progress curve (AUDPC) values and relative area under disease progress curve (rAUDPC) (Madden et al. 2007). AUDPC was calculated as:

AUDPC=Σni=1((yi+yi+1/2)(ti+1 - ti))AUDPC=Σi=1n((yi+yi+1/2)(ti+1 - ti))

where y_i = rust severity at the i-th observation, t_i = time (days) at the i-th observation, and n = total number of observations. The rAUDPC was calculated as:

rAUDPC=AUDPC/(tf - t0)×100rAUDPC=AUDPC/(tf - t0)×100

where t_f = the duration of days at the final rating and t_0 = the time of disease onset. Percent disease control (hereafter referred to as disease control) was calculated at each location using the formula ((1 – (rAUDPC of a treatment / rAUDPC of the NTC) × 100). Yield was collected from the center two plot rows at plant maturity. Yield loss assessment was evaluated by correlating R7 leaf severities to percent change in yield from the nontreated control. Only trials where observed rust severity on the NTC was greater than 5% at R7 and where significant yield differences occurred were used for yield loss assessment. Namely, CREC 2008 Timing, CREC 2009 Efficacy, CREC 2009 Timing, and LREC 2010.

Fungicide efficacy

Rust developed at all locations in 2008, and R7 severity was high enough on the NTC at CREC 2008 and CASS 2008 to see differences in disease control among treatments. Due to the low amount of disease pressure at LREC 2008, statistically significant levels of disease control were not found among the fungicides tested and numerically negative disease control values were an artifact of very low disease pressure. At CREC 2008 and CASS 2008, disease control provided by DMI fungicides was statistically higher than that of the only SDHI tested at that location, penthiopyrad. At CREC 2008, statistical differences between the QoI treatments pyraclostrobin and azoxystrobin were observed. At CASS 2008, rust severity at R5 and R7 was relatively low, but year-end severities were among the highest at physiological maturity

In 2009, rust developed at all locations with great enough severity that disease control differences among treatments were observed. DMI fungicide applications resulted in greater levels of disease control than a boscalid application at locations where it was tested. Greater disease control was achieved with tebuconazole and prothioconazole + tebuconazole applications than azoxystrobin at CREC and GRAN in 2009. At LREC 2009, levels of disease control did not significantly differ between applications of pyraclostrobin and the DMI fungicides tested. At CREC 2009, applications of DMIs and pyraclostrobin resulted in a statistically greater yield than the NTC. However, yield between the azoxystrobin treatment and the NTC was the same.

Fungicide timing—Normal onset

Rust developed at all locations in 2008, but severity was great enough only at CASS 2008 and CREC 2008 to observe meaningful differences in disease control. Yield differences were observed only at CREC 2008. At CASS 2008, disease control from fungicide applications ranged from 97.3 to 71.6%, and statistical differences were observed among treatments. Yield was not statistically different than the NTC for any fungicide application. At CREC 2008, significant differences in disease control and yield were observed among fungicide timings. Levels of disease control and yield were the same between a single application of pyraclostrobin at R5 and three fungicide applications of pyraclostrobin, both of which were statistically different than the NTC.

Rust developed at all locations in 2009, and severity was great enough to see meaningful statistical difference in disease control. In CREC 2009 and LREC 2009, disease control with an R5 application of pyraclostrobin or tebuconazole was statistically the same as three applications of fungicides. Statistical differences in disease control were observed among single application timings in both trials. Five fungicide treatments in CREC 2009 resulted in statistically higher yield than the NTC, and differences in yield among treatments was observed. At GRAN 2009, rust was not effectively controlled by singular applications made at R6.

Fungicide timing—Early onset

An early disease onset was not achieved at any locations in 2010; however, rust severities were high enough to observe significant differences among treatments. At CREC 2010, disease control was significantly higher than the NTC when single applications of pyraclostrobin were made. Also, disease control levels were the same with pyraclostrobin applications made at R1, R5, and V8-V12 + R1 + R5. No differences were observed among treatments made up of multiple applications of pyraclostrobin. At GRAN 2010, disease control from single fungicide applications at V8-V12 and R1 was statistically the same as the NTC. At LREC 2010, disease control levels of four fungicide treatments were significantly different than the NTC. Disease control values were among the highest for treatments containing multiple applications of pyraclostrobin when a R5 timing application was included. However, yield differences were not observed among treatments with multiple applications.

An early disease onset in 2011 did not occur at any location. However, significant differences in disease control were observed among treatments. At GRAN 2011, disease control was greater than the NTC for all applications except a single application at V8-V12 of pyraclostrobin. The highest levels of disease control were achieved with multiple applications of a fungicide, which often included a R5 application. At LREC 2011, disease control was significantly higher for all fungicide applications than the NTC. Levels of disease control and yield were statistically the same among multiple pyraclostrobin applications at V8-V12 and R1, R1 and R5, and at all three timings.

Yield loss assessment.

To determine the impact of rust on yield, trials CREC 2008-Normal Onset, CREC 2009 Efficacy, CREC 2009-Normal Onset, and LREC 2010 were selected for analysis because R7 severity values on the NTC were greater than 5% and significant differences were observed in yields. A negative correlation of r = -0.7756 was found between percent change in yield from the NTC and fungicide treatment at R7 severity values. Specifically, for every 1% increase in severity at R7, yields decreased by 6.6%.

Fungicide timing proved to be a critical component in disease management. Of all single timings evaluated throughout this study, disease reductions appeared most frequently with a single application at R5. Applications at earlier or later growth stages tended to control disease less than applications made at R5. This could be because applications made earlier than R5 may have not provided protection to the top leaves prior to leaf expansion, or that the duration of protection was inadequate for the length of growing season remaining. Conversely, an application later than R5 may have had a more limited impact on disease control because a high severity had been reached by the time of application. In four trials, disease control from a single R5 fungicide application reduced rust as much as three applications (CREC 2008-Normal Onset, CREC 2009-Normal Onset, GRAN 2009-Normal Onset, and LREC 2009-Normal Onset). With the exception of GRAN 2009, it is notable that these trials had the highest R7 rust severities of the six 2008 and 2009 timing trials. This indicates that reduction of rust with a single application was as effective as three applications, under the highest disease pressure situations observed in this study.

Although one objective of the research project was to evaluate fungicide application timing in an "early onset" disease scenario in 2010 and 2011, rust did not develop early enough to simulate an early onset epidemic. However, some trends were noticed. Based on disease control values, single applications made at V8-V12 did not provide satisfactory disease control and multiple applications of a fungicide that included an R5 application often were among the highest for disease control. Prophylactic fungicide applications for yield gain independent of disease control have been suggested on sunflower at the R1 growth stage. Trials established in 2010 and 2011 included a R1 application, but no correlation between single R1 applications and yield was observed.

Yield loss correlations with sunflower rust severity are very limited in the literature. This is largely because sunflowers are sensitive to numerous yield-limiting and unmanageable problems such as bird damage, insects, other diseases, and high winds. Similar stresses were observed in our study; for example, sunflower midge caused a total yield loss at CREC 2010, Phomopsis stem canker reduced yields by 50% in LREC 2011, and straight-line winds (estimated to be between 26.8 and 33.5 ms^{-1}) caused 100% lodging at GRAN 2010. Consequently, yield loss data under sunflower rust pressure in the literature is limited. Of the 17 trials in our study, eight trials had rust pressure in excess of 5% at R7, four of which had statistical differences yield differences that were believed to be primarily or solely due to rust. In these trials, a negative correlation ($r = -0.7756$) was observed between R7 severity and percent change in yield, resulting in a yield reduction of 6.6% for every 1% increase in rust severity.

Based on previous work, an action threshold of 3% severity was recommended to reduce disease pressure and limit yield loss. Data from these studies indicate that an application at a lower disease severity is more appropriate, but choice of DMI or QoI product infrequently matters. As a result of these studies, a threshold of 1% disease severity on the upper four leaves, when occurring at or before R5, has been suggested and widely accepted in the United States by the sunflower industry and producers.

TOBACCO

Tobacco, common name of the plant *Nicotiana tabacum* and, to a limited extent, Aztec tobacco (*N. rustica*) and the cured leaf that is used, usually after aging and processing in various ways, for smoking, chewing, snuffing, and extraction of nicotine. Various other species in the genus *Nicotiana* are grown as ornamentals, known collectively as flowering tobaccos.

Origin of the Crop

The information available on the origin and history of tobacco goes generally from ambiguity and contradictions. Plant geneticists have established that tobacco's centre of origin i.e the meeting place between a species' genetic origin and the area in which it was first cultivated, is located in the Peruvian/Ecuadoraen Andes. Estimates for its first date of cultivation range from 5000-3000 BC. According to one source, tobacco was in existence in Asia even during the 12th century, when it was not known elsewhere. It was not only used as an intoxicant but also as a cure for all kinds of ills and paying homage to deities. However, it was Christopher Columbus who discovered the narcotic qualities of tobacco by accident in the course of his American voyage in 1492. On landing in the Islands of Tobago, Columbus and his men were taken by surprise to find the natives either sniffing a powdered dry leaf with evident pleasure or smoking roughly made roll of dried-up leaves. On trying these themselves, Columbus and his men were satisfied with the intoxicant produced. They took along with them some quantity of dried leaves as well as that of the seeds and that was how tobacco got introduced into Europe.

The Red Indians according to another version, used tobacco for both medicinal and ceremonial purposes. They used to inhale its smoke from burning leaf through the nostrils by means of a hollow forked cane and the name of the instrument was given to the plant which came to be known as 'Tobaco' in Spanish and 'Tobacco' in English. The plant was first introduced into Europe in the year 1560 by a Spanish physician sent to Mexico. About this time Jean Nicot, the French Ambassador to Portugal came to know of tobacco in Lisbon and introduced it to the French Court. The botanical name of the plant Nicotiana and the word nicotine have been derived from his name. The habit of smoking spread in several countries during the 17th century.

Tobacco is said to have been introduced into India in the beginning of 17th century. Tobacco, as elsewhere in the world, has thrived in spite of considerable neglect and social disapproval in India. Currently it is an important commercial crop of the country earning around 6000 crores foreign exchange and 20000 crores excise revenue. Further, it is providing livelihood security to 45 mi people including farmers, farm laborers, tendu leaf puckers, bidi rollers, traders etc.

Modes of use

There is a wide range of use of tobacco in different countries.

The smoking form of tobacco, since its introduction in South Asian countries, has been used in several forms, like hukka (water pipe), chilam (clay pipe), cigarettes, rolled tobacco in the form of bidees, Chchuta (reverse smoking), etc., whereas the nonsmoking or chewable tobacco is in the form of snuff/naswar (roasted and finely powdered for inhalation), mawa, qiwam, gutkha, kheni (mixture of dry raw tobacco with lime), zarda, betel quid with tobacco, paan-masala, etc.. In Indonesia, tobacco is mixed with clove and dipped in the oral cavity.

Paan acquired significant popularity among the male population of the central and western India in contrast to female of these locations. Contrary are the findings in Bangladesh, with educated or uneducated females consuming more paan preferably with tobacco. According to a survey report, tobacco-consuming females in Bangladesh believe that this helps them in concentrating more on their work. The survey also states that a large population believes that paan strengthens their teeth and eliminates bad breath.

Cultivation

Though tobacco is tropical in origin, it is grown throughout the world. Cultivated tobacco (*N. tabacum*) requires a frost-free period of 100 to 130 days from date of transplanting to maturity in the field. Aztec tobacco (*N. rustica*), which is grown to some extent in India, Vietnam, and certain Transcaucasian countries, matures more quickly and is more potent than cultivated tobacco.

The prime requisite for successful tobacco culture is a supply of well-developed healthy seedlings that is available at the proper time for transplanting. Soil for a plant bed should be fertile and of good tilth and drainage; it must be protected from chilling winds and exposed to the sun. The soil is usually partially sterilized by burning or using chemicals such as methyl bromide (now illegal in many countries) to control plant diseases, weeds, insect pests, and nematodes.

The soil must be finely pulverized and level so that the seed can be lightly covered with soil by rolling or trampling. Uniform distribution of seeds is important. In warm regions of the world, the germinating seedlings are produced outdoors in cold frames covered with thin cotton cloth or a thin mulch, such as chopped grass (used in particular in Zimbabwe), straw, or pine needles. Glass or plastic is used in colder regions, and close attention is given to watering and ventilation. After 8 to 10 weeks the seedlings are 10 to 18 cm (4 to 7 inches) in length and are ready for transplanting in the field. Transplanting machines are used extensively in some areas, but much of the world's tobacco is planted by hand.

Spacing of plants in the field varies widely according to the type of tobacco. Orinoco strains, used for flue curing, are grown in rows 1.2 metres (4 feet) apart, with plants 50 to 60 cm (20 to 24 inches) apart in the row. Varieties in the Pryor group are grown to produce the dark air-cured and fire-cured types and are often planted in hills 1 metre (3.5 feet) apart. Burley and Maryland strains, used for the production of light air-cured tobaccos, may be planted 81 to 91 cm (32 to 36 inches) apart or closer. Broadleaf and seed-leaf strains, including the Havana seed, Cuban, and Sumatra varieties, are used for the production of cigars; they are grown in rows spaced 1 metre (3 feet) apart, with individual plants placed at a distance of 38 to 68 cm (15 to 27 inches) from each other. The variety grown for production of Perique is spaced the widest, with rows 1.5 metres (5 feet) apart and 91 to 107 cm (36 to 42 inches) between plants. Aromatic tobaccos, also used for cigars, are spaced in rows 38 to 60 cm (15 to 24 inches) apart with 8 to 20 cm (3 to 8 inches) between plants in the row.

Soil requirements vary widely with the type of tobacco grown, though well-drained soil with good aeration is generally desirable. Flue-cured, Maryland, cigar-binder, and wrapper types of tobacco are produced on sandy and sandy loam soil. Burley, dark air-cured, fire-cured, and cigar-filler types are grown on silt loam and clay loam soils, with clay subsoils. The need for fertilizer is determined by the type of tobacco, soil, and climate; nitrogen, phosphorus, and potassium are commonly applied as necessary to prevent symptoms of nutritional deficiency.

Fungicides to Manage Diseases

Fungicides are organized into groups based on their mode of action, which is the mechanism by which a pesticide or chemical affects its target pest. Fungicide groupings are determined by an agrochemical industry group known as the Fungicide Resistance Action Committee (FRAC), and the groupings are referred to as FRAC Groups or FRAC Codes. While fungicides may have different product names, if they are in the same group, they essentially cause the same challenge to the fungus.

How Resistance to Fungicides Develops

When a fungicide with a particular mode of action is first introduced, most target fungi in the environment are very sensitive to it, but there may be one or two individuals that are naturally resistant, or even just slightly less sensitive to the fungicide mode of action. As this mode of action is used repeatedly, most sensitive target fungi are suppressed, but the naturally resistant fungi are less affected. These resistant fungi contribute relatively more descendants to the next pathogen generation, gradually building up a population that overall is less sensitive to the fungicide. After many applications of the same mode of action, the population becomes composed of more and more descendants of resistant fungi, and fewer and fewer descendants of sensitive fungi, which results in poorer disease control when using that mode of action.

Fungicide Resistance Management

Management of resistance to fungicides is based on alternating the use of particular modes of action, or FRAC groups, which essentially presents multiple different challenges to the fungal population. Overall, fungi that are naturally resistant to a mode of action are very rare in the environment. Challenging a population with multiple different modes of action will reduce the chance of developing widespread resistance, which will prolong the efficacy of these chemicals.

There have been many recent advances in fungicide chemistries since the early 1980s, and tobacco growers may now choose from seven modes of action for foliar disease management.

At a minimum, growers are encouraged to alternate FRAC codes to maintain fungicide efficacy. However, when only two FRAC codes are in use, there still may be potential to develop resistance to both modes of action, especially if the fungicides used are prone to resistance development. Proneness for development of resistance is based on whether fungicides have single-site or multi-site modes of action.

Fungicides with single-site modes of action are effective on a very specific part of the fungal lifecycle, which reduces effects on nontarget organisms, and these often have systemic activity, meaning the chemical actually enters the plant tissue and may have some curative activity if present at infection.

Unfortunately, these fungicides are more prone to development of resistance in pathogen populations, and include fungicides in groups 11, 43, and 40. Fungicides with multi-site modes of action, such as those in group M3, are effective on multiple points of the pest lifecycle, and are less prone to development of resistance, yet many of these are protectant fungicides with no systemic activity. An exception is Actigard, which induces the plant's own defenses against infection by pathogens, resulting in systemic activity yet a low potential for resistance.

Blue Mold Management

Several products are listed for the 'sample program choice,' with the single best choice based on individual farm choices. For example, growers whose last application in the floathouse was Quadris (to manage target spot) would be ill-advised to use any azoxy fungicide as their first field application, for concerns about developing resistance to the azoxystrobins.

However, a grower with frogeye pressure would want to incorporate an azoxystrobin product by the second application, to address both blue mold and frogeye disease concerns. Other fungicides, such as Manzate ProStick, are not recommended close to harvest as a result of concerns about residues and the mandatory preharvest. The preharvest interval, which is found on the product label, should be adhered to for all fungicides used.

Diseases in Seed Beds

Damping off : Pythium Aphanidermatum

This is a fungal disease caused by the soil borne pathogen. It is the most common and serious disease in tobacco nurseries causing death of seedlings. Conspicuous symptom of this disease is the sudden collapse of young seedlings in patches leading to uneven stand. Brown watery soft rot of young seedlings, girdling of hypocotyls and finally toppling and death of seedlings leading to wet rot are the characteristic symptoms. The pathogen spreads quickly and affect the seed bed causing enormous loss of seedlings.

High humidity, high soil moisture, cloudiness, temperature below 24oC, continuous wet weather, location of nursery in low lying areas are the favourable factors for high incidence of damping off disease.

Cultural and Preventive Measures

- Deep ploughing in summer destroys the fungus.
- Preparation of raised seed beds 15 cm high with channels around to provide drainage.
- Rabbing the seed bed before sowing with slow burning farm waste materials like paddy husk, tobacco stubbles, waste grass etc.
- Seed rate @ 3.5 kg/ha is to be used to avoid over crowding of seedlings.
- Waterings are regulated to avoid excessive dampness on bed surface.

Chemical control

- Bordeaux mixture @ 0.4% (40 gm of copper sulphate + 40 gm of lime in 10 litres of water) or Fytolan or Blitox @ 0.2% (20 gm in 10 litres of water) is to be applied with rose can 2 weeks after sowing. This has to be repeated once in 4 days under normal weather conditions and once in 2 days under wet and cloudy conditions. In case of heavy rain, the application has to be repeated.
- Spraying ridomil MZ 72 W.P. @ 20 g or fenamodone 10%+ mancozeb 50% @ 30 g in 10 litres of water twice at 20 and 30 days after germination. Ridomil should not be sprayed before 20 days after germination or more than two times or more than the recommended dose. (500 g of ridomil is sufficient for one acre of nursery).

Black shank and leaf blight : Phytophthora parasitica var. nicotianae

This is a soil borne disease. It frequently occurs during mid nursery period causing leaf blight and blackening of roots and stems leading to death of seedlings. Water soaked brown to black lesions appear on the leaf. These patches enlarge and coalesce leading to wet rot of leaf tissue and midribs. Cloudy weather, prolonged dampness and temperature below 22oC are the favourable factors for sudden outbreak of this disease.

Cultural and preventive measures

Black shank can be controlled by following the same schedule suggested for 'Damping-off'. Mc.Nair-12 or CM-12(KA) or K-326 which are black shank tolerant varieties are to be used in irrigated northern light soils.

Chemical control

Spraying Fytolan or Blitox @ 2gin 10 concentration when the seedlings are 50-60 days old controls leaf blight. Ridomil MZ 72 W.P. @ 20 g or Fenamidone + mancozeb 60 % @ 30g in 10 litres of water can also be used to control this disease.

Collar rot : Sclerotium rolfsii

Just like damping off, sudden death of seedlings in patches is noticed in seed beds. Blackening of the collar region, wilting and rotting of leaves are the symptoms.

Control measures

Spraying 0.4% Bordeaux mixture or Ziram 20 g or Blitox 20 g in 10 litres of water 3 weeks after seed germination. This is to be sprayed twice or thrice at weekly intervals depending upon weather conditions.

Leaf spot diseases

Anthracnose : Colletotrichum tabacum

This disease can be seen in any stage of the nursery. It will cause heavy loss to the nursery when favourable conditions to the fungus prevail. Small, light green to white water soaked lesions develop on young leaves. These lesions enlarge in wet weather to form oily, circular spots of 3 mm diameter. The spots dry up, become papery, thin, grey white surrounded by brown border. Effected leaves become wrinkled and distorted. Elongated, black or brown lesions on the midrib and petiole are seen in severe cases. On the stem, several elongated or oblong sunken brown lesions develop making the seedlings unfit for transplanting.

Optimum temperature for the occurrence of this disease is 18oC. High relative humidity, reduced light and overcast weather are favourable for the outbreak of this disease.

Frog-eye spot : Cercospora nicotianae

Generally this disease is seen 4-5 weeks after germination. Brown, round spots resembling frog eye form appear on the lower leaves of the seedlings. Spots appear first on basal leaves and gradually spread to upper leaves. Spots are small, circular, brown or tan with grey centres. In severe cases coalesce to become bigger spots leading to drying up of leaves.

Frequent waterings and wet weather leading to high humidity and temperature around 27°C are favourable for the development of the disease.

Control Measures

Spraying with Carbendazim 50 WP 3 gm or pyraclostrobin 5% + metiram 55% WG @ 20 g or Azosystrobin 23% SC @ 0.1% in 10 litres of water at 30 days and 40 days after germination controls this disease. If necessary another spray has to be given after 10 days.

Root-knot nematode: Meloidogyne javanica/ Meloidogyne incognita

The light soil nurseries are prone to this disease. Root knot is caused by nematodes and prevalent in most of the light soil nurseries. Sandy soils and red sandy loams with adequate soil moisture favour the disease development. Several weeds growing in and around nurseries are alternate hosts of root knot nematodes. They are responsible for build-up of the nematodes in the soil.

Characteristic symptoms of root knot disease are yellowing of leaves, stunted growth of seedlings, wilting of plants, premature death of seedlings leading to bald patches in seed beds. Affected seedlings when pulled out show several galls on the roots. Galls vary in size, coalesce and produce multiple galls which give the seedlings a sickly appearance. Root knot affected seedlings do not establish well in the field resulting in heavy gaps.

Control Measures

- Soon after the nursery season, ploughing of the nursery area thoroughly destroys the nematode population by exposing them to the hot sun. The area should be ploughed subsequently, 3 or 4 times during summer months also.
- Before raising the nursery, the soil must be tested for nematodes.
- Rabbing the seed beds before sowing with paddy husk @ 6 kg/m2 or slow burning farm waste material to destroy nematode larvae and egg masses gives initial protection to seedlings for about a month.
- Soil solarisation by covering the nursery soil with 100 gauge white alkathene sheet during summer for about 6-8 weeks controls nematodes effectively.
- Nursery should not be grown where brinjal or tomato were grown previously.
- Rotation of the nursery area with crops resistant to nematodes like groundnut, redgram, marigold, cotton, gingelly and chillies is recommended. If such rotation crops are grown for more than three years the nematodes can be controlled.
- The site of nursery should be changed every year. If one year the nursery is infected with nematodes, nursery should not be grown in that site next year.
- Use of tray seedlings fortified with P.lilacinus (50 g) or P. chlamydosporia (50 g).

Tobacco Mosaic Virus : Mormor tabaci

This is a contagious disease caused by virus. The infested leaves show characteristic mottling with light and dark green patches and seedlings are stunted.

Control Measures

- Workers should disinfect their hands with soap and running water before weeding or handling seedlings.
- Workers should not chew or smoke while at work in nursery.
- Diseased seedlings should be removed and destroyed.
- Phyto sanitary measures are very important for control of the disease.

Diseases in field crop

Black shank : Phytophthora parasitica f. nicotianae

This is a soil borne disease and it occurs when there is heavy and continuous rainfall, high soil moisture and relative humidity. The blackening of the stem of the diseased plants is noticed upto 30 cm height from the base. At the blackened portion inside the stem, the pith transforms into black round plate like discs. Roots also will be blackened, becomes soft and finally rottening is seen. The leaves of the effected plants becomes yellow, wilt and finally plant dies.

Control Measures

- In fields known for severe infection with Black shank, Mc.Nair-12 or CM-12 (KA) or Kanchan varieties which are tolerant to black shank are to be used in northern light soils.
- Any one of the following fungicides @ 100 ml per plant is to be applied in planting hole and then planting is done.
- Bordeaux mixture 0.4% (40 gm of Copper sulphate + 40 gm of lime in 10 litres of water).
- Copper oxychloride (Blitox or Fytolan) @ 0.2% (20 gm in 10 litres of water).

Leaf Spot Diseases

Frog eye spot : Cercospora nicotianae

The causal agents, symptoms and favourable weather conditions are the same as described under nursery diseases. In addition, this disease also appears as minute dark brown or black spots on stems, midribs, flower stalk and capsules.

Control Measures

Spraying Carbendazim 50 WP @ 3 gm or pyraclostrobin 5% + metiram 55% WG @ 20 g or Azosystrobin 23% SC @ 0.1% in 10 litres of water) at 15 days interval, commencing from 30 days after planting, effectively controls frog eye spot disease.

Brown spot : Alternaria alternata

Brown spot is a major disease affecting both FCV and non-FCV tobacco's in India. It is prevalent in most parts of India. This disease generally occurs when there is high relative humidity and atmospheric temperatures around 20-25oC. Especially when dew fall is more, more number of spots will be seen on the leaves. Dark brown or burnt spots separately or coalesced, appear on the diseased leaves. The spots will be like small water soaked spots first, later gradually grow to one cm size in 5-6 days and turns to brown colour. Sometimes yellow rings are seen around the spots just like frog eye spot. When the spots are closely observed, there will be circular rings on the spots. The disease will first appear on the lower leaves and gradually spreads to the upper leaves. Affected leaves dry up and do not cure properly. Dark brown, sunken, elongated spots appear on stems, petioles, seed capsules and stalks also.

Control Measures

Mancozeb @ 20 g in 10 litres of water in combination with plantamycin or streptomycin 3 gm or propiconazol 10ml / 10 liters of water is to be sprayed on the crop such that all the leaves are covered. For this high volume sprayers like sikar or hi-tech or knapsack sprayers should be used. Depending on the intensity of the disease, two or three sprays should be given at weekly or 10 days interval. Use of excess doses of nitrogen fertilisers should be avoided.

Broomrape (Orobanche) : Orobanche cernua

It is a complete root parasite affecting the yield and quality of tobacco. The shoots emerge in clusters and their basal portion is attached to tobacco roots through which it draws nourishment and depletes the host resulting in yield loss of 24 to 52%. Affected plants become stunted, leaves turn pale and wilt. Initially leaf tips droop and as the attack intensifies, all the leaves wilt. High soil moisture due to irrigation or rain after planting, low soil temperature during winter months encourage heavy incidence of Orobanche.

Control Measures

- Deep ploughing twice or thrice in summer buries Orobanche seed to deeper depth and there by helps in reducing the emergence of Orobanche.
- In Orobanche sick fields, growing tobacco for one or two seasons is to be skipped off.
- Avoiding growing of brinjal, tomato and bhendi crops in the sick fields.
- Periodical removal of Orobanche shoots before flowering and setting of seeds reduces the menace. The Orobanche shoots should be destroyed by burning.
- Growing trap crops such as jowar, gingelly, blackgram and greengram in kharif facilitates orobanche germination but will not allow it to grow. This reduces the Orobanche seed load in the soil.
- CTRI developed a spear with a 2 m stick at the end of which 18 cm length, 8 cm breadth and 0.5 cm thick sharp iron blade is arranged. Cutting the Orobanche shoots within 3-4 days after emergence out of the soil, before flowering of Orobanche either upto the soil level or 2-3 cm below the soil level with spear effectively controls orobanche.

Leaf curl virus: Ruga tabaci

This virus is transmitted through whitefly. Different types of viruses cause different symptoms. With mild types the leaves show slight wrinkles and drooping. In case of severe types the leaves and stem curls and leaves become thick and shrinkled and unfit for curing. Sometimes enations are also seen on under side of the upper leaves.

Diseased leaves show vein clearing, puckering of leaves, downward curling of leaf margins; leaves become brittle; thickening of veins. Since whiteflies spread disease, conditions such as too much bushy vegetation around tobacco fields serve as breeding places for whiteflies and thus spread the disease. Growing of brinjal, sunflower nearby tobacco fields also encourage build-up of whiteflies.

Control Measures

- Alternate weed hosts around nursery area are removed and destroyed.
- Yellow-sticky traps (20 cm x 15 cm size galvanized iron sheet painted with yellow colour coated with castor oil) are installed @ 5 per acre.
- If the population of whitefly is 100 per each sticky trap, the following insecticides may be sprayed at 10 days interval commencing from 4 weeks after germination.
- Imidacloprid 200 S.L. @ 2.5 ml in 10 litres of water (or)
- Thiamethoxam 25 WG @ 2gm (or) pymetrozine 50 WG @ 4g (or) flonicamid 50 WG @ 4g in 10 lit of water

Spraying has to be done preferably in the evening hours (6-4 pm) with high volume sprayers and it should be ensured that the both sides of the leaves also are covered. Further the spraying schedule has to be adopted on community basis.

Tobacco Mosaic Virus: Mormor tabaci

The leaves infested with tobacco mosaic virus show dark or light green patches. This is a contagious disease. This disease spreads through contact by labour and implements used for interculture. If the plants are infected in the early stages they show stunted growth reducing the yield and quality considerably. If the disease occurs when the crop is full grown, the tobacco yields are not affected.

Unclean cultivation, indiscriminately touching diseased and healthy plants, use of tobacco products by workers while working in field, presence of susceptible weeds and crop plants near fields are some of the factors responsible for disease development.

Control Measures

- In areas where TMV is a serious problem resistant varieties like CTRI Sulakshana, VT 1158 are recommended.
- Since it is a contagious disease, the following sanitary measures are suggested to prevent the disease spread.
- Washing hands in soap water before and after field operations, roguing the diseased plants early in the season and avoiding smoking during field operations.
- Prophylactic sprays with virus inhibitors of plant origin like Basella alba and Bougainvillea spectabilis and neem leaf extracts @ 1% dilution on 30th, 40th and 50th day of planting tobacco are useful.
- Mosaic affected plants are to be removed within 3 weeks after planting and replaced with healthy seedlings.

GLOSSARY

Abrasive – Capable of wearing away or grinding down another object. Some formulations, such as wettable powders, wear out spray nozzles more quickly, necessitating more frequent calibration to ensure that rates are not exceeded.

Absorbed Dose – In exposure assessment, the amount of a substance that penetrates an exposed organism's absorption barriers (skin, lung tissue, gastrointestinal tract wall, etc.) through physical or biological processes. The term is synonymous with internal dose.

Absorption – The uptake of pesticides into plants and animals. Absorption of pesticides by target and nontarget organisms is influenced by environmental conditions and by the chemical and physical properties of the pesticide and the soil. Once absorbed by plants, pesticides may be broken down or they may remain in the plant until tissue decay or harvest.

Absorption Barrier – Any of the exchange sites of the body that permit uptake of various substances at different rates (e.g., skin, lung tissue, gastrointestinal tract wall, etc.)

Acid Deposition – A complex chemical and atmospheric phenomenon that occurs when emissions of sulfur and nitrogen compounds and other substances are transformed by chemical processes in the atmosphere, often far from the original sources, and then deposited on earth in either wet or dry form. The wet forms, popularly called "acid rain," can fall to earth as rain, snow, or fog. The dry forms are acidic gases or particulates.

Acid Equivalent (a.e.) – The theoretical yield of parent acid from the active ingredient of a herbicide that has been formulated as a derivative of an acid. Rates of phenoxy herbicides, for example, are expressed in terms of acid equivalent.

Action Levels – (1) Regulatory levels recommended by EPA for enforcement by FDA and USDA when pesticide residues occur in food or feed commodities for reasons other than the direct application of the pesticide. As opposed to "tolerances" which are established for residues occurring as a direct result of proper usage, action levels are set for inadvertent residues resulting from previous legal use or accidental contamination. (2) In the Superfund program, the existence of a contaminant concentration in the environment high enough to warrant action or trigger a response under SARA and the National Oil and Hazardous Substances Contingency Plan.

Activated Carbon/Activated Charcoal – A highly adsorbent form of carbon used to remove odors and toxic substances from liquid or gaseous emissions. In waste treatment, it is used to remove dissolved organic matter from waste drinking water. In drinking water treatment plants, it is used to remove taste and odor as well as various contaminants.

Active Ingredient (a.i.) – The active chemical(s) in a pesticide formulation.

Acute Toxicity – A measure of the quantity of a chemical, as a single dosage or concentration, required to cause injury or illness of test animals.

Adjuvant – Chemical added to a pesticide formulation or to the spray tank to increase its effectiveness, safety, or application characteristics.

Adsorption – A process that binds pesticides to soil particles, often because of the attraction between the chemical and soil particles. Positively charged pesticide molecules, for example, are attracted to and can bind to negatively charged clay particles. Soils high in organic matter or clay are more adsorptive than coarse, sandy soils, in part because a clay or organic soil has more particle surface area, or more sites onto which pesticides can bind. Wet soils tend to adsorb less pesticide than dry soils because water molecules compete with the pesticide for the binding sites. Some pesticides such as paraquat and glyphosate bind very tightly, while others bind only weakly and are readily desorbed or released back into the soil solution. Some pesticide labels recommend higher application rates when the chemical is applied to adsorptive soils, because of reduced availability. However, injury can result when a pesticide used for one crop is later released from the soil particles in amounts great enough to cause injury to a sensitive rotational crop. Adsorption is particularly important because it influences whether other processes are able to affect pesticides; for example, soil microorganisms cannot degrade pesticides while they are adsorbed.

Adulterated – (1) Any pesticide whose strength or purity falls below the quality stated on its label. (2) A food, feed, or product that contains illegal pesticide residues.

Adverse Effect – An unexpected adverse condition (such as an unreasonable health effect or unexpected damage to nontarget crops or aquatic/terrestrial organisms) or a new finding in the product chemistry and manufacture of the technical material, which was not previously identified by the registrant. An adverse effect report must be submitted to the EPA by the registrant each time an adverse effect is found.

Aerated Lagoon – A holding and/or treatment pond that speeds up the natural process of biological decomposition of organic waste by stimulating the growth and activity of bacteria that degrade organic waste.

Aerial Application – Pesticide treatment applied with the use of an airplane or helicopter.

A-Horizon – Surface and subsurface with most of the organic matter, microbial and chemical breakdown reactions, and plant roots; varies in depth depending on the geological origins and degree of erosion.

Air Pollutants – Substances in the air that interfere with human health or welfare, or produce other harmful environmental effects; common categories are solids, sulfur compounds, volatile organic chemicals, particulate matter, nitrogen compounds, oxygen compounds, halogen compounds, radioactive compound, and odors.

Alkaline – The opposite of acidic; having a pH greater than 7. Some pesticides are adversely affected in alkaline spray solutions or carryover longer in high pH soils.

Allelopathy – The adverse effect on the growth of plants or microorganisms caused by the action of chemicals produced by other living or decaying plants.

Allergic Effects – Harmful effects, such as skin rash or asthma, that some people develop in reaction to pesticides but that do not cause the same reaction in most other people.

Annual – A plant that completes its life cycle from seed in less than one year.

Antidote (Protectant, Safener) – (1) A chemical added to a pesticide (usually a herbicide) or applied to the crop (usually the seed), to prevent or reduce the harmful effects of a pesticide on a crop. (2) An antidote is also a substance used as a medical treatment to counteract poisoning.

Aquifer, Unconfined – An aquifer containing water that is not under pressure; the water level in the associated well is the same as the water table outside the well. Conversely, confined aquifers are under pressure.

Bacillus thuringiensis (Bt) – A naturally occurring bacterial disease of insects. Insecticides have been manufactured that contain this bacteria as the active ingredient. Bacillus thuringiensis insecticides are most commonly used against certain leaf and needle-feeding caterpillars. Recently, strains of the bacteria have been produced that affect certain fly larvae, such as mosquitoes, and larvae of leaf beetles. Bacillus thuringiensis is considered safe to humans and non-target species, such as wildlife. Some formulations can be used on essentially all food crops.

Back-Siphoning – The movement of liquid pesticide mixture back through the filling hose and into the water source.

Bacteria (Singular – bacterium) – Microscopic living organisms that can aid in pollution control by metabolizing organic matter in sewage, spill sites, or other contaminated areas. However, there are also bacteria in soil, water or air that can cause human, animal and plant health problems.

Band Application – An application of a pesticide to a continuous narrow strip, usually over, along, or in a crop row, rather than broadcast over the entire field area. Banding reduces the pesticide load on the environment. However, under some circumstances, such as coincidence of ammonia injection furrows and pesticide bands, banding may increase movement of the pesticide through the soil. Also, the additional cultivation which is often needed for weed control between the bands can make herbicide banding unacceptable in areas where cultivation is being avoided to reduce erosion.

Basal Treatment – A pesticide application to the stems or trunks of woody plants at and just above the ground.

Bed – A narrow, flat-topped ridge on which crops are grown, with a furrow on each side for irrigation or drainage of excess water. The bed may be rebuilt during the season as it washes away.

Best Management Practices (BMPs) – Soil, nutrient, and pesticide conservation practices that also provide water quality benefits. They include numerous practices, such as cover crops, green manure crops, and strip-cropping to control erosion, soil testing and timing of chemical applications to prevent the loss of nutrients and pesticides, use of adjuvants to maximize pesticide effectiveness at lowest rates, and careful selection of application methods such as banding to reduce rates.

B-Horizon – Subsoil where leached substances are broken down very slowly and are thus more likely to reach groundwater.

Biennial – A plant that completes its life cycle after an overwintering period which is necessary to initiate the flowering response.

Bioassay – Use of living organisms to determine the presence of a pesticide.

Bioavailability – Degree of ability to be absorbed and ready to interact in organism metabolism.

Bioconcentration – The accumulation of a chemical in tissues of a fish or other organism to levels greater than in the surrounding medium.

Biodegradable – Capable of decomposing under natural conditions.

Biodiversity – Refers to the variety and variability among living organisms and the ecological complexes in which they occur. These items can be organized at many levels, ranging from complete ecosystems to the biochemical structures that are the molecular basis of heredity. Thus, the term encompasses different ecosystems, species, and genes.

Biological Carrying Capacity – The maximum number of animals of a given population that an environment can sustain.

Biological Control/Biocontrol – Man's use of a specially chosen predator, parasite, or disease to attack a harmful insect, plant pathogen, or weed.

Biological Magnification – Refers to the process whereby certain substances such as pesticides or heavy metals move up the food chain, work their way into rivers or lakes, and are eaten by aquatic organisms such as fish, which in turn are eaten by large birds, animals or humans. The substances become concentrated in tissues or internal organs as they move up the chain.

Biological/Biochemical Oxygen Demand (BOD) – An indirect measure of the concentration of biologically degradable material present in organic wastes. It usually reflects the amount of oxygen consumed in five days by biological processes breaking down organic waste. The greater the BOD, the greater the degree of pollution.

Bioremediation – Use of living organisms to clean up oil spills or remove other pollutants from soil, water, or wastewater; use of organisms such as non-harmful insects to remove agricultural pests or counteract diseases of trees, plants, and garden soil.

Biotechnology – Techniques that use living organisms or parts of organisms to 1) produce a variety of products (from medicines to industrial enzymes) to improve plants or animals, 2) develop microorganisms to remove toxins from water or soil, or 3) act as pesticides.

Bloom – A proliferation of algae and/or higher aquatic plants in a body of water; often related to pollution, especially when pollutants accelerate growth.

Breakdown of Pesticides – The degradation of pesticides, usually beneficial. Pesticide-destroying reactions change most pesticide residues in the environment to nontoxic or harmless compounds. However, degradation is detrimental when a pesticide is destroyed before the target pest has been controlled. See Chemical Breakdown, Microbial Breakdown, and Photodecomposition.

Broadcast Application – An application of a pesticide uniformly to soil or plants; the term foliar application is usually preferred when insecticides or fungicides are applied to plants.

Broadleaf Weeds – In general, dicotyledonous weeds with net-veined leaves and terminal or axillary meristems.

Buffer Strips – Areas or strips of land maintained in permanent vegetation, designed to intercept pollutants and erosion. Placed around fields, they can enhance wildlife habitat, improve water quality, and enrich aesthetics on farmlands. Various types of buffers include contour buffer strips, filter strips, riparian forest buffers, field borders, windbreaks/shelterbelts, hedgerows, grassed waterways, and alley cropping.

Carcinogen – A substance or agent capable of producing cancer in animals.

Carrier – A gas, liquid, or solid substance used to dilute or suspend a pesticide during its application.

Carryover – When a pesticide remains in the soil in sufficient quantity to detrimentally affect the following crop(s). There is more potential for carryover in soils high in clay, organic matter, CEC, and pH (high pH reduces chemical and microbial breakdown while increasing availability of some pesticides), and less potential in soils high in microbial and chemical activity (which are favored by a warm, moist, well-aerated, fertile soil with a medium soil pH). Certain pesticide characteristics (low water solubility, strong adsorption, and low susceptibility to microbial or chemical degradation) increase potential for carryover. Carryover potential can be reduced by uniformly applying the lowest recommended rate, accurate acreage determination and chemical measurement, and proper sprayer calibration. Banding of pesticides, early-season application, tillage and cultivation, tank mixtures to reduce the rate of the persistent partner, and selection of a tolerant rotational crop or variety help minimize carryover problems.

Chemical Breakdown – The degradation of pesticides by processes that do not involve living organisms. Temperature, moisture, pH and adsorption, in addition to the chemical and physical properties of the pesticide, determine which chemical reactions take place and how quickly they occur. One of the most common pesticide degradation reactions is hydrolysis, a breakdown process in which the pesticide reacts with water in the spray mix or soil. Many organophosphate and carbamate insecticides are particularly susceptible to hydrolysis under alkaline conditions. Some are actually broken down within a matter of hours when mixed with alkaline water. Product labels may warn against mixing a pesticide with certain fertilizers, other pesticides or water with specific characteristics. Following these precautions can help prevent pesticide degradation and potential incompatibility problems in the spray tank. In some situations, buffers or other additives may be available to modify spray mix conditions and prevent or reduce degradation. Pesticide degradation and possible corrosion of application equipment can be avoided by not allowing a spray mix to remain in a tank for a long period of time.

Chemical Name – The systematic name of a chemical compound according to specific rules of nomenclature. The accepted chemical name is usually the name used in Chemical Abstracts.

Chemical-Resistant – Able to prevent movement of the pesticide through the material during the period of use.

Chemigation – The application of pesticides through irrigation water. The chemigation unit must be calibrated with each use to ensure accurate application. Use a secondary containment structure made of impermeable material where pesticides are stored near the irrigation well when chemigation is practiced, in case of a leak or spill. Minimize runoff and leaching by adjusting amounts and timing based on soil moisture and crop needs to maintain vigorous growth, avoiding movement beyond root zone.

Chemnet – Mutual aid network of chemical shippers and contractors that assigns a contracted emergency response company to provide technical support if a representative of the firm whose chemicals are involved in an incident is not readily available.

Chemtrec – The industry-sponsored Chemical Transportation Emergency Center; provides information and/or emergency assistance to emergency responders.

Chlorosis – Loss of green color (chlorophyll) in foliage, resulting in yellowing of the plant.

Chronic Exposure – Multiple exposures occurring over an extended period of time or over a significant fraction of an animal's or human's lifetime (usually seven years to a lifetime).

Chronic Toxicity – A measure of the quantity of a chemical required to cause injury or illness of test animals, after lifetime exposure.

Clean Water Act – The 1977 Clean Water Act established the basic structure for regulating discharges of pollutants into the waters of the United States. It gave EPA the authority to implement pollution control programs such as setting wastewater standards for industry, made it unlawful to discharge any pollutant from a point source into navigable waters (unless a permit was obtained under its provisions), funded the construction of sewage treatment plants and recognized the need for planning to address the critical problems posed by nonpoint source pollution.

Code of Federal Regulations (CFR) – Document that codifies all rules of the executive departments and agencies of the federal government. It is divided into fifty volumes, known as titles. Title 40 of the CFR (referenced as 40 CFR) lists all environmental regulations.

Collecting Pad or Tray – A safety system designed to contain and recover spills, rinsates, leaks, and other pesticide-containing substances.

Common Name – A non-Latin name for a plant species, or a simple (not Chemical Abstracts) name used to identify the active ingredient of a pesticide.

Community Water System (CWS) – A public water system which serves at least 15 service connections used by year-round residents, or regularly serves at least 25 year-round residents. See Non-Community Water System.

Companion Crop/Nurse Crop – A crop (commonly a cereal) grown with another crop (commonly a small-seeded legume) in order to secure a return from the land in the first year of a new seeding of the slow-growing legume. The companion crop also reduces the potential for erosion during the establishment of the legume.

Composting – The controlled biological decomposition of organic material in the presence of air to form a humus-like material. Controlled methods of composting include mechanical mixing and aerating, ventilating the materials by dropping them through a vertical series of aerated chambers, or placing the compost in piles out in the open air and mixing or turning it periodically.

Concentrate – Applied in reduced water volume compared to what is normal for that crop/size/density. Concentrate sprays are often applied in orchards. Also see Dilute.

Concentrated Flow – Runoff that accumulates or converges into well-defined channels.

Confidential Business Information (CBI) – Material that contains trade secrets or commercial or financial information that has been claimed as confidential by its source (e.g., a pesticide or new chemical formulation registrant). EPA has special procedures for handling such information.

Confidential Statement of Formula (CSF) – A list of the ingredients in a new pesticide or chemical formulation. The list is submitted at the time for application for registration or change in formulation.

Conservation – The use, protection, renewing, and improvement of natural resources according to principles that will ensure their highest economic or social benefits.

Contact Herbicide – A herbicide that kills primarily the plant tissue to which it is applied. Such a herbicide will kill seedlings or young plants, but established perennial plants generally recover because below-ground parts are not injured.

Contour Buffer Strips – Series of strips of grass or legumes placed across the slope on a contour, which help trap sediment and nutrients. This is similar to strip-cropping, but with narrower grass or legume strips. The alternating strips of grass or other permanent vegetation slow runoff flow, trap sediment from the crop strips above, and increase water infiltration. Because the buffer strip is established on the contour, runoff flows evenly across the entire surface of the grass strip, reducing sheet and rill erosion.

Contour Furrows – Furrows plowed nearly level around the hill – at right angles to the slope. Crop row ridges subsequently built by tilling and/or planting on the contour create hundreds of small dams. These ridges or dams slow water flow and increase infiltration which reduces erosion.

Contour Strip-Cropping – Crop rotation and contouring combined in equal-width strips of row crop (typically corn or soybeans) planted on the contour and alternated with strips of oats, grass, or legumes. Not more than half a field can be planted to row crops. Oats, grass, or legume slows runoff, increases infiltration, traps sediment and provides surface cover. Ridges formed by contoured rows slow water flow thus reducing erosion. Rotating the strips from corn to legumes allows nutrient-needy crops to benefit from the nitrogen added to the soil by legumes.

Council for Agricultural Science and Technology – A nonprofit organization composed of scientific societies and many individual, student, company, nonprofit, and associate society members, which assembles, interprets, and communicates science-based information regionally, nationally, and internationally on food, fiber, agricultural, natural resource, and related societal and environmental issues, for various stakeholders-legislators, regulators, policy makers, the media, the private sector, and the public.

Cover Crop– A crop grown between crops or between seasons, which is not necessarily harvested but is grown to cover the soil and thus reduce erosion.

Cover Spray – An application after petal fall of fruiting trees, when leaves are fully-expanded.

Critical Area Planting – Planting grass, legumes, trees or shrubs in small, isolated areas of excessive erosion. The vegetation provides surface cover to stop the raindrop splash and slow water flow.

CropLife America – A trade association representing most of the manufacturers of pesticides.

Crop Residue Management – A year-round system beginning with the selection of crops that produces sufficient quantities of residue and may include the use of cover crops after low residue producing crops. CRM includes all field operations that affect residue amounts, orientation and distribution throughout the period requiring protection. Leaving last year's crop residue on the surface before and during planting operations by reducing tillage provides cover for the soil at a critical time of the year, shielding soil particles from rain and wind until plants can produce a protective canopy. Site-specific residue cover amounts needed are usually expressed in percentage but may also be in pounds. See Tillage, Conservation for tillage systems used with CRM.

Crop Rotation – Changing the crops grown in a field from year to year in a planned sequence. Crop rotations can reduce weed, disease, insect and other pest problems by changing the environment or interrupting their food source; provide alternative sources of soil nitrogen (because alfalfa and other legumes replace some of the nitrogen that corn and other grain crops remove); reduce soil erosion; and reduce risk of water contamination by agricultural chemicals. Crop rotation is a common practice on sloping soils because of its potential for soil saving.

Cultivation – Tillage operations performed for weed control after seeding the crop, using equipment such as a rolling cultivator, conventional row crop cultivator, rotary hoe, or spike-tooth harrow.

Cultural Carrying Capacity – The number of animals that a human or human community will tolerate in a given area.

Cultural Weed Control – Providing the crop with the optimum growing conditions, and utilizing genotype, crop rotation, plant population, row spacing, competitiveness, and other plant-related characteristics so that the crop is able to compete more successfully with weeds.

Data Call-In – A part of the Office of Pesticide Programs (OPP) process of developing key required test data, especially on the long-term, chronic effects of existing pesticides, in advance of scheduled Registration Standard reviews. Data Call-In from manufacturers is an adjunct of the Registration Standards program intended to expedite re-registration.

Data Compensation – Under US law, companies who are the original submitters of data to support a product registration, are entitled to receive financial compensation for the data when their competitors rely on such data to obtain their own or substantially similar pesticide registrations (e.g. generic registrations, or me-too registrations). For a new active ingredient, FIFRA 3(c)(1)(F) entitles the original data submitter to 10 years exclusive use protection plus 5 years of data compensation. For all other eligible data, 15 years of data compensation is allowed from the date of the submission.

Decontamination – The removal of pesticide from surfaces or organisms that were exposed, so no further harm or damage can occur.

Defoliant – A chemical that causes the leaves to abscise from a plant.

Delayed Dormant – An application when a plant, typically a fruiting tree, is still dormant but about to enter the pre-bloom stage.

Dermal Exposure – Contact between a chemical and the skin.

Desiccant – Any substance or mixture of substances used to accelerate the drying of plant tissue.

Detection Limit – The lowest concentration of a chemical that can reliably be distinguished from a zero concentration.

Detention Time – (1) The theoretical calculated time required for a small amount of water to pass through a tank at a given rate of flow. (2) The actual time that a small amount of water is in a settling basin, flocculating basin, or rapid-mix chamber of a water treatment plant. (3) In storage reservoirs, the length of time water will be held before being used.

Development Effects – Adverse effects such as altered growth, structural abnormality, functional deficiency, or death observed in a developing organism.

Dilute – Applied in standard water volume for that crop/size/density. Also see Concentrate.

Directed Application – An application of a fungicide or insecticide to a specific part of the crop; or an application of a herbicide to weeds while contacting no more than the lower part of the stems of row crops. This application is done to improve pest control and, with some herbicides, to limit the potential for crop injury.

Directed Band – An application in a continuous narrow strip while contacting no more than the lower part of the stems of row crops.

Direct Filtration – A method of treating water which consists of the addition of coagulant chemicals, flash mixing, coagulation, minimal flocculation, and filtration. Sedimentation is not used.

Disease Cycle – The successive stages from primary infection through incubation, appearance of symptoms, and production of secondary inoculum, which can start the cycle again.

Disposal – Final placement or destruction of toxic, radioactive, or other wastes; surplus or banned pesticides or other chemicals; polluted soils; and drums containing hazardous materials from removal actions or accidental releases. Disposal may be accomplished through use of approved secure landfills, surface impoundments, land farming, deep-well injection, ocean dumping, or incineration. See Pesticides, Proper Handling.

Dissolved Oxygen (DO) – The oxygen freely available in water, vital to fish and other aquatic life and for the prevention of odors. DO levels are considered a most important indicator of a water body's ability to support desirable aquatic life. Secondary and advanced waste treatment steps are generally designed to ensure adequate DO in waste-receiving waters.

Dissolved Solids – Disintegrated organic and inorganic material in water. Excessive amounts make water unfit to drink or use in industrial processes.

Diversion – A channel or earthen embankment, much like a terrace, constructed across the land slope to intercept and divert surface runoff from a specific area to an outlet. A diversion is often built at the base of a slope to divert runoff away from bottom lands. A diversion may also be used to divert runoff flows away from a feedlot, or to collect and direct water to a pond.

Dormancy – The state of inhibited seed germination or growth of a plant. Winter annuals, biennials, and perennial plants usually have a period of dormancy during the winter. Dormancy may be genetically or environmentally induced.

Dose-Response – Shifts in toxicological responses of an individual (such as alterations in severity) or populations (such as alterations in incidence) that are related to changes in the dose of any given substance.

Double Cropping – Two crops grown in quick succession, instead of the normal practice of growing one or the other. Most commonly, winter wheat is harvested and immediately followed by a late-seeded soybean variety that requires a slightly shorter growing season than a conventional variety.

Drag-Off – Cultivation (by harrowing) of the hill before the crop (especially potatoes) emerges. Herbicides may be applied after drag-off for control of unemerged weeds.

Drainage, Subsurface – The removal of excess water from the soil profile by means of drain tiles, perforated pipes, or other devices.

Drainage, Surface – The diversion or orderly removal of excess water from the surface of the land by means of improved natural or constructed channels, supplemented when necessary by the sloping and grading of land surfaces to these channels.

Drench Application – A pesticide poured around the base of the tree or on the tree trunk.

Drift – Movement of airborne drops of spray solution, or vapors, from the intended area of application to nontarget species. Generally granules and pellets reduce drift compared to wettable powders and other liquid sprays. Dusts are most susceptible to drift. For liquid sprays, adjust equipment to minimize fine droplets, which are most susceptible to drift. Release pesticide spray as close to the target as possible, in calm (10 mph or less) weather conditions. Windy conditions or air conditions created by a temperature inversion (cold air trapped between the soil surface and warm air above) generally contribute to pesticide drift.

Drinking Water State Revolving Fund – Fund which provides capitalization grants to states to develop drinking water revolving loan funds to help finance system infrastructure improvements, assure source-water protection, enhance operation and management of drinking-water systems, and otherwise promote local water-system compliance and protection of public health.

Drip Application – Slow, monitored release of pesticide through drip irrigation equipment to conserve water.

Dust – A formulation providing excellent coverage of mature crops with dense foliage, because its carrier (talc, clay, finely ground plant parts, etc.) has very small particle sizes. It is highly susceptible to drift.

Dystrophic Lakes – Acidic, shallow bodies of water that contain much humus and/or other organic matter; they contain many plants but few fish. See eutrophic lakes.

Early Preplant (EPP) – Herbicides applied two weeks or more before planting the crop. Products that can be applied early preplant as well as preplant and preemergence give the grower a wider application window, making adverse weather conditions like rainy weather less likely to hinder or prevent the application. Stewardship is also aided by products that give growers the flexibility to wait instead of spraying prior to adverse weather conditions such as heavy rain.

Ecological Risk Assessment – The application of a formal framework, analytical process, or model to estimate the effects of human action(s) on a natural resource and to interpret the significance of those effects in light of the uncertainties identified in each component of the assessment process. Such analysis includes initial hazard identification, exposure and dose-response assessments, and risk characterization.

Ecosystem – An ecological entity consisting of the biotic community and the nonliving environment functioning together in an interacting system.

Emergence – The breaking through the soil surface by a plant seedling or an elongating shoot.

Emulsifier – Chemical that allows petroleum-based pesticides (EC's) to mix with water.

Emulsion – A mixture of two or more liquids that are not soluble in one another. One is suspended as small droplets in the other.

Endangered Species – Organisms whose survival as a species has been designated by a Federal Agency to be endangered or threatened.

Endangered Species Act – This 1973 Act provides a program for the conservation of threatened and endangered plants and animals and the habitats in which they are found. The US Fish and Wildlife Service maintains the list of endangered and threatened species. Species include birds, insects, fish, reptiles, mammals, crustaceans, and plants ranging from grasses to trees. Anyone can petition FWS to include a species on this list. The law prohibits any action, administrative or real, that results in a "taking" of a listed species, or adversely affects habitat. Likewise, import, export, interstate, and foreign commerce of listed species are all prohibited. EPA's decision to register a pesticide is based in part on the risk of adverse effects on endangered species as well as environmental fate (how a pesticide will affect habitat). Under FIFRA, EPA can issue emergency suspensions of certain pesticides to cancel or restrict their use if an endangered species will be adversely affected. EPA, FWS, and USDA are preparing county bulletins that include habitat maps, pesticide use eliminations, and other actions required to protect listed species.

Enrichment – The addition of nutrients (e.g., nitrogen, phosphorus, carbon compounds) from sewage effluent or agricultural runoff to surface water, greatly increasing the growth potential for algae and other aquatic plants.

Environment – The sum total of all biological and physical factors affecting an organism, population, or community.

Environmental Protection Agency – The government agency formed in 1970 and charged with protecting human health and the environment. EPA works to develop and enforce regulations that implement environmental laws enacted by Congress. EPA delegates to states and tribes the responsibility for issuing permits and for monitoring and enforcing compliance. Where national standards are not met, EPA can issue sanctions and take other steps to assist the states and tribes in reaching the desired levels of environmental quality

Epinasty – That state in which more rapid growth on the upper side of a plant organ or part (especially leaf) causes it to bend or curl downward.

Erosion – The detachment and movement of surface soil and rock particles by gravity, wind, water, freezing and thawing, and/or other natural phenomena, which is intensified by land-clearing practices related to farming, residential or industrial development, road building, or logging.

Eutrophication – The slow aging process during which a lake, estuary, or bay evolves into a bog or marsh and eventually disappears. During the later stages of eutrophication the water body is choked by abundant plant life due to higher levels of nutritive compounds such as nitrogen and phosphorus. Human activities can accelerate the process.

Eutrophic Lakes – Shallow, murky bodies of water with concentrations of plant nutrients causing excessive production of algae. See dystrophic lakes.

Exemption – An excuse given by a state (with primacy) to a public water system from a requirement involving a Maximum Contaminant Level (MCL), treatment technique, or both, if the system cannot comply due to compelling economic or other factors, or because the system was in operation before the requirement or MCL was instituted; and the exemption will not create a public health risk. Also called a variance.

Experimental Use Permit (EUP) – A permit granted by EPA that allows a producer to conduct tests of a new pesticide, product and/or use outside the laboratory. The testing is usually done on ten or more acres of land or water surface, as opposed to small research plots.

Eyewash Dispenser – Commercially available system for flushing contaminants out of the eyes.

Fallow – Cropland left idle, usually for one growing season in dryland areas, during which time either herbicides (chemical fallow) or tillage is used to control weeds and build up moisture reserves for the next crop.

Fall Preplant – Herbicides applied in the fall, prior to planting the crop the following spring. Timing is critical relative to soil freeze-up, to ensure that product does not degrade or move in the environment during the fall or winter months.

Farm Pond – A water body typically formed by building a dam across an existing gully or low-lying area. Earth for the dam is dug out above the dam with heavy machinery to form a bowl, to supply water for livestock, recreation and wildlife, and to control gully erosion. Generally the ponded area fills with water within a year. An overflow pipe is installed through the dam to control the water level and allow water to spill through the dam without causing erosion.

Field Border – Strips of perennial vegetation (grass or legumes) established at the outside edges of a field where excessive sheet and rill erosion is occurring. The grass or legume strips replace crop end rows, which would be planted up and down hill and be highly erosive. Field borders are sometimes referred to as picture frames of grass, and are used with contour farming, terrace, buffer strip and contour strip-cropping systems. The grass or legume in the strip protects steep field edges from soil erosion, and provides turning and travel lanes around the field.

Filter Strip – A strip of grass, trees, or shrubs that filters runoff and removes sediment, fertilizer, and pesticides before they reach water bodies or water sources including wells. Strips of grass, trees and/or shrubs slow water flow and cause contaminants like sediment, pesticides, and fertilizers to collect in vegetation. Collected nutrients are used by the vegetation, rather than entering water supplies. Filtered water then enters water bodies.

Finished Water – Water that has passed through all the processes in a water treatment plant and is ready to be delivered to consumers.

Foliar Application – A uniform application of a pesticide to emerged plants. Also see Broadcast Application.

Food and Drug Administration – One of our nation's oldest consumer protection agencies, charged with monitoring the manufacture, import, transport, storage and sale of about 1$ trillion worth of products each year. Foremost, FDA is a public health agency, charged with protecting American consumers by enforcing the Federal Food, Drug, and Cosmetic Act and several related public health laws. To carry out this mandate of consumer protection, FDA has some 1,100 investigators and inspectors who cover the country's almost 95,000 FDA-regulated businesses. Food (e.g. pesticide residues), cosmetics, medicines, microwave ovens, and food and drugs for pets are among the myriad of products under FDA jurisdiction.

Formulation – A physical mixture of one or more pesticides plus non-pesticide ingredients by a manufacturer for practical use. The formulation may influence effectiveness, selectivity, ease of application, packaging, and cost.

Fumigant – Pesticide that is a vapor or gas or that forms a vapor or gas when applied and whose pesticidal action occurs in the gaseous state.

Fumigation – A pesticide application in gaseous or volatile liquid form, usually made in an enclosed area or incorporated into the soil.

Fungi – (Singular – fungus) Molds, mildews, yeasts, mushrooms, etc., which lack chlorophyll (are not photosynthetic) and which are usually non-mobile, filamentous, and multicellular. Some grow in soil while others attach themselves to decaying trees and other plants to obtain nutrients. Some are pathogens while others are beneficial, such as those used to stabilize sewage and digest composted waste.

Genetic Engineering – A process of inserting new genetic information into existing cells in order to modify a specific organism for the purpose of changing one of its characteristics.

Germination – The process of initiating growth in seeds, spores, or other plant propagules.

Inoculum – The pathogen or its parts that can cause infection. That portion of individual pathogens that are brought into contact with the host.

Integrated Pest Management (IPM) – An ecosystem-based, multi-disciplinary strategy using biological, chemical, cultural, mechanical, genetic, and educational methods to keep pests below unacceptable economic thresholds. IPM includes pest prevention, regular pest and weather monitoring, habitat manipulation, crop rotation, natural enemies, biological control, pest-resistant varieties, pest attractants and repellents, biopesticides, and synthetic organic pesticides. In its best form, it is a long-term strategy which minimizes any detrimental effects on human health, the environment, and non-target organisms.

Juvenile Hormones – Natural insect chemicals that keep the earlier stages of an insect from changing into the normal adult form.

LC50 – The lethal concentration of a chemical that will kill %50 of the test organisms within a designated period. The lower the LC50, the more toxic the compound.

LD50 – The lethal dose of a chemical that will kill %50 of the test organisms within a designated period. The lower the LD50, the more toxic the compound.

Leaching – The downward movement of pesticides in water moving through the soil profile, rather than over the surface (the latter is runoff). Leaching depends, in part, on the pesticide's chemical and physical properties. For example, a pesticide held strongly to soil particles by adsorption is less likely to leach. Another factor is solubility. A pesticide that dissolves in water can move with water in the soil. The persistence, or longevity, of a pesticide also influences the likelihood of leaching. A pesticide that is rapidly broken down by a degradation process is less likely to leach because it may remain in the soil only a short time. Soil factors that influence leaching include texture and organic matter, in part because of their effect on pesticide adsorption. Soil permeability (how readily water moves through the soil) is also important. The more permeable a soil, the greater potential for pesticide leaching. A sandy soil is much more permeable than a clay. The method and rate of application, the use of tillage systems that modify soil conditions, and the amount and timing of water a treated area receives after application can also influence pesticide leaching. Typically, the closer the time of application to a heavy or sustained rainfall, the greater the likelihood that some pesticide leaching will occur. A certain, small amount of pesticide leaching may be essential for control of a target pest. Too much leaching, however, can lead to reduced pest control, injury of nontarget species and groundwater contamination. Monitoring weather conditions and the amount and timing of irrigation can help minimize pesticide leaching. Careful pesticide selection is important because those pesticides that are not readily adsorbed, not rapidly degraded, and highly water soluble are the most likely to leach. The label may also advise against using the pesticide when certain soil, geologic or climatic conditions are present. Pesticides can leach through the soil to groundwater from storage, mixing, equipment cleaning, disposal areas, and even from normal applications. It is critical to know the geology and the relative depth of the groundwater in your area to prevent contamination by leaching.

Necrosis – Localized death of tissue usually characterized by browning and desiccation.

Nematodes – Small, usually microscopic, eel-like roundworms.

Nitrogen Fixation – Conversion of nitrogen in air to nitrogen compounds which plants can use, by soil organisms or organisms living in nodules on legume roots.

Organic Farming – A production system, which avoids or largely excludes the use of synthetic fertilizers, pesticides, growth regulators, and livestock feed additives. To the maximum extent feasible, organic farming systems rely upon such techniques as crop rotations, crop residues, animal manures, legumes, green manures, mechanical cultivation, and aspects of biological pest control to maintain soil productivity and tilth, to supply plant nutrients, and to control insects, weeds and other pests.

Organic Matter – Plant and animal residues at various stages of decomposition, soil organisms, and substances synthesized by them.

Pathogen – An organism that causes disease in other organisms.

Restricted-Use Pesticide – A pesticide that is available for purchase and use only by certified pesticide applicators or persons under their direct supervision, because it requires special handling due to a relatively high potential for human, environmental, or other hazards.

Rinsate – Pesticide-containing water (or another liquid) that results from rinsing a pesticide container, pesticide equipment, or other pesticide-containing materials.

Soil Application – Application of an herbicide made primarily to the soil surface rather than to vegetation.

Soil Conditioner – An organic material like humus or compost that helps soil absorb water, build a bacterial community, and take up mineral nutrients.

Soil Erodibility – An indicator of a soil's susceptibility to raindrop impact, runoff, and other erosive processes.

Soil Incorporation – Mechanical mixing or leaching in with water of a pesticide into the surface soil to facilitate uptake by pests or crop, or to reduce its loss by volatility or photodecomposition.

Soil Residual Life – The length of time that a pesticide remains active in the soil.

Soil Residue – When a pesticide remains in the soil after the target crop has been harvested.

Soil Texture – Relative proportion of sand, silt, and clay in the soil. Soils with high sensitivity to groundwater contamination have a coarse texture (high proportion of sand) and relatively low relief.

Soil Water Zone – The zone extending from the land surface down through the major root zone, capable of holding water. Therefore, its total depth is variable and is dependent upon soil type and vegetation. This zone is unsaturated except during periods of heavy infiltration and percolation.

Spray Solution – What is applied to the field when a pesticide formulation is added to a liquid carrier (water, liquid fertilizer, oil, etc.) in the spray tank. The spray "solution" is actually a true solution, an emulsion, or a suspension, depending on whether the pesticide formulation dissolves, disperses, or is suspended in the carrier. Emulsions and suspensions can settle without proper agitation before and during application.

Tailings – Residue of raw material or waste separated out during the processing of crops or mineral ores.

Tail Water – The runoff of irrigation water from the lower end of an irrigated field.

Tillage – Mechanical soil stirring operations carried on prior to seeding the crop. Various types of equipment are used for tillage, depending on the objective of the tillage. The type and quality of tillage has important environmental impacts. The goal is always optimum pulverization and partial residue coverage because too little or too much will increase run-off. The goal of the final tillage operation (seedbed preparation) is optimum soil particle size reduction to allow close soil contact with the seed while maintaining pores for water and air movement, and a rough surface that dries rapidly and won't germinate weeds. Cultivation is the preferred term after seeding the crop, though the terms are often used synonymously.

Water-Based Pesticides – Pesticides that use water as the only diluent or carrier.

Water Quality – The chemical, physical, biological, and radiological condition of a surface or ground water body.

Water Quality Criteria – Levels of water quality expected to render a body of water suitable for its designated use. Criteria are based on specific levels of pollutants that would make the water harmful if used for drinking, swimming, farming, fish production, or industrial processes.

Watershed – The area of land draining to a specific water outlet (stream, reservoir, etc.); the boundary is defined by the region's topography. Watersheds vary in size and can be nested within other larger watersheds. Land use within a watershed largely determines the quality of the local surface water. The quality of water leaving a watershed can, in turn, affect the cumulative quality of water far downstream. For example, pesticides detected in a city's drinking water supply could come from lawn and other urban uses or from an upstream watershed where agriculture is predominant.

Water Storage Pond/Water Treatment Lagoon – An impound for liquid wastes designed to accomplish some degree of biochemical treatment.

Water Table– The upper level of the groundwater in an area. The water table level fluctuates throughout the year, lowering as water is removed from wells or discharged at streams and springs. The water table rises through recharge from rain and melting snow that seeps through soil into the aquifer. For years it was believed that the natural filtering of water during its slow movement through the soil, sand, gravel and rock formations was adequate to cleanse it of contaminants before it reached groundwater. Today, many chemicals, including some pesticides, have been detected in groundwater. Studies have shown that recharge can carry pollutants down to aquifers. Furthermore, it is clear that human activities can lead to contamination of the recharge water.

Zero Tolerance – Indicates that no detectable amount of the specified pesticide may remain on the raw agricultural commodity when it is sold or utilized.

BIBLIOGRAPHY

Aladjadjiyan A. Physical Factors for Plant Growth Stimulation Improve Food Quality. In: Aladjadjiyan, A., editor. Food Production - Approaches, Challenges and Tasks. Rijeka (Croatia): InTech; 2012. p. 145-168

Anderson, A. J., Tari, P. H., and Tepper, C. S. 1988. Genetic studies on the role of an agglutinin in root colonization by *Pseudomonas putida*. Appl. Environ. Microbiol. 54: 375-380.

Andrews, J. 1992. Biological control in the phyllosphere. Annu. Rev. Phytopathol. 30: 603-633.

Anonymous, *Headline Fungicide Supplemental Label*, BASF Corporation, Florham Park, NJ, USA, 2008.

Audenaert, K., Pattery, T., Cornelis, P., and Hofte, M. 2002. Induction of systemic resistance to *Botrytis cinerea* in tomato by *Pseudomonas aeruginosa* 7NSK2: role of salicylic acid, pyochelin and pyocyanin. Mol. Plant-Microbe Interact. 15: 1147-1156.

Balba, H., "Review of strobilurin fungicide chemicals," *Journal of Environmental Science and Health - Part B Pesticides, Food Contaminants, and Agricultural Wastes*, vol. 42, no. 4, pp. 441–451, 2007.

Bargabus, R. L., Zidack, N. K., Sherwood, J. W., and Jacobsen, B. J. 2002. Characterization of systemic resistance in sugar beet elicited by a non-pathogenic, phyllosphere colonizing *Bacillus mycoides*, biological control agent. Physiol. Mol. Plant Pathol. 61: 289-298.

Bargabus, R. L., Zidack, N. K., Sherwood, J. W., and Jacobsen, B. J. 2004. Screening for the identification of potential biological control agents that induce systemic acquired resistance in sugar beet. Biological Contr. 30: 342-350.

Bartlett, D. W., J. M. Clough, J. R. Godwin, A. A. Hall, M. Hamer, and B. Parr-Dobrzanski, "The strobilurin fungicides," *Pest Management Science*, vol. 58, no. 7, pp. 649–662, 2002.

Benhamou, N. 2004. Potential of the mycoparasite, *Verticillium lecanii*, to protect citrus fruit against *Penicillium digitatum*, the causal agent of green mold: A comparison with the effect of chitosan. Phytopathology 94: 693-705.

Benhamou, N., and Chet, I. 1997. Cellular and molecular mechanisms involved in the intersection between *Trichoderma harzianum* and *Pythium ultimum*. Appl. Environ. Microbiol. 63: 2095–2099.

Bhagwat S. Haytowitz DB, Holden JM. USDA Database for the Isoflavone Content of Selected Foods, Release 2.0. U.S. Department of Agriculture, Agricultural Research Service, NutrientData Laboratory. http: //www.ars.usda.gov/nutrientdata/isoflav

Biermann, B., and Linderman, R. G. 1983. Use of vesicular-arbuscular mycorrhizal roots, intraradical vesicles and extraradical vesicles as inoculum. New Phytol. 95: 97-105.

Bradley, C. A., "Frogeye leaf spot pathogen with reduced sensitivity to fungicides found in Tennessee soybean field," *University Illinois Extension Bulletin* 172, 2010.

Brown-Rytlewski, D. and P. Vincelli, *Letter from Universities Regarding the Strobilurin, Pyraclostrobin (Headline), Supplemental Label*, U.S. Environmental Protection Agency, Washington, DC, USA, 2009.

Bull, C. T., Shetty, K. G., and Subbarao, K. V. 2002. Interactions between Myxobacteria, plant pathogenic fungi, and biocontrol agents. Plant Dis. 86: 889-896.

Burks, C., Darby, A., Londono, L. G., Momany, M., and Brewer, M. T. 2021. Azole-resistant Aspergillus fumigatus in the environment: Identifying key reservoirs and hotspots of antifungal resistance. PLOS Pathogens.

Catska, V. 1994. Interrelationship between vesicular-arbuscular mycorrhiza and rhizosphere microflora in apple replant disease. Biologia Plant. 36: 99-104.

Chisholm, S. T., Coaker, G., Day, B., and Staskawicz, B. J. 2006. Host-microbe interactions: shaping the evolution of the plant immune response. Cell 124: 803-814.

Cober ER, Cianzio SR, Pantalone VR. Soybean. In: Vollmann, J. and Rajcan I. editors. Oil Crops. Springer-Verlag New York Inc, 2009. p. 57-9

Corn Disease Working Group. 2022. Fungicide efficacy for control of corn diseases. Corn Disease Management. CPN-2011.

Corn Disease Working Group. Mueller, D., et al. 2021. Corn disease loss estimated from the United States and Ontario Canada - 2020. CPN-2007-20. Published March 18, 2021.

Culbreath, A. K., T. B. Brenneman, and R. C. Kemerait, "Management of early leaf spot of peanut with pyraclostrobin as affected by rate and spray interval," *Plant Health Progress*, 2002.

De Meyer, G., and Hofte, M. 1997. Salicylic acid produced by the rhizobacterium *Pseudomonas aeruginosa* 7NSK2 induces resistance to leaf infection by *Botrytis cinerea* on bean. Phytopathology 87: 588-593.

Duchesne, L. C. 1994. Role of ectomycorrhizal fungi in biocontrol. Pages 27-45 in: Mycorrhizae and Plant Health. F. L. Pfleger and R. G. Linderman, eds. APS Press, St. Paul, MN.

Duijff, B. J., Gianinazzi-Pearson, V., and Lemanceau, P. 1997. Involvement of the outer-membrane lipopolysaccharides in the endophytic colonization of tomato roots by biocontrol *Pseudomonas fluorescens* WCS417r. New Phytol. 135: 325-334.

Elad, Y., and Baker, R. 1985. Influence of trace amounts of cations and siderophore-producing pseudomonads on chlamydospore germination of *Fusarium oxysporum* Ecol. Epidemiol. 75: 1047-1052.

Fitter, A. H., and Garbaye, J. 1994. Interactions between mycorrhizal fungi and other soil microorganisms. Plant Soil 159: 123-132.

Fry, W. E. 1982. Principles of Plant Disease Management. Academic Press. New York, NY.

Garcia-Garrido, J. M., and Ocampo, J. A. 1989. Effect of VA mycorrhizal infection of tomato on damage caused by *Pseudomonas syringae*. Soil Biol. Biochem. 21: 165-167.

Glandorf, D. C., Verheggen, P., Jansen, T., Jorritsma, J. W., Smit, E., Leefang, P., Wernars, K., Thomashow, L. S., Laureijs, E., Thomas-Oates, J. E., Bakker, P. A., and Van Loon, L. C. 2001. Effect of genetically modified *Pseudomonas putida* WCS358r on the fungal rhizosphere microflora of field-grown wheat. Appl. Environ. Microbiol. 67: 3371-3378.

Gould, F. and C. Gautreaux, "Odds are low for grain sorghum disease, but still scout," LSU AgCenter, 2015.

Green, M. B. and D.A. Spilker. 1986. Fungicide Chemistry: Advances and Practical Applications (ACS Symposium Series, 304). American Chemical Society. Oxford University Press.

Hagan, A. K., H. L. Campbell, K. L. Bowen, and L. Wells, "Impact of application rate and treatment interval on the efficacy of pyraclostrobin in fungicide programs for the control of early leaf spot and southern stem rot on peanut," *Peanut Science*, vol. 30, no. 1, pp. 27–34, 2003.

Harman, G. E., Howell, C. R., Vitarbo, A., Chet, I., and Lorito, M. 2004. *Trichoderma* species - opportunistic, avirulent plant symbionts. Nature Rev. Microbiol. 2: 43-56.

He, P., Chintamanani, S., Chen, Z., Zhu, L., Kunkel, B. N., Alfano, J. R., Tang, X., and Zhou, J. M. 2004. Activation of a COI1-dependent pathway in *Arabidopsis* by *Pseudomonas syringae* type III effectors and coronatine. Plant J. 37: 589-602.

Hemingway J, Eskandari M, Rajcan I. Genetic and Environmental Effects on Fatty Acid Composition in Soybeans with Potential Use in Automotive Industry. Crop Science. 2015, 51: 1-11

Henry, R. S., W. G. Johnson, and K. A. Wise, "The impact of a fungicide and an insecticide on soybean growth, yield, and profitability," *Crop Protection*, vol. 30, no. 12, pp. 1629–1634, 2011.

Hoitink, H. A. J., and Boehm, M. J. 1999. Biocontrol within the context of soil microbial communities: a substrate dependent phenomenon. Annu. Rev. Phytopathol. 37: 427-446.

Hollier, C., "Commercial crop production field crops-rain sorghum," Louisiana Field Crops, 2015.

Homma, Y., Kato, Z., Hirayama, F., Konno, K., Shirahama, H., and Suzui, T. 1989. Production of antibiotics by *Pseudomonas cepacia* as an agent for biological control of soilborne plant pathogens. Soil Biol. Biochem. 21: 723-728.

Howell, C. R., and Stipanovic, R. D. 1980. Suppression of *Pythium ultimum* induced damping - off of cotton seedlings by *Pseudomonas fluorescens* and its antibiotic, pyoluterin. Phytopathology 70: 712-715.

Howell, C. R., Beier, R. C., and Stipanovic, R. D. 1988. Production of ammonia by *Enterobacter cloacae* and its possible role in the biological control of *Pythium* pre-emergence damping-off by the bacterium. Phytopathology 78: 1075-1078.

Hutson, D. and Miyamoto, J. 1999. Fungicidal Activity: Chemical and Biological Approaches to Plant Protection. John Wiley & Sons. New York, NY.

Hymowitz T, Collins FI. Variability of sugar content in seed of Glycine max (L.) Merrill. and G.soja Sieb. and Zucc. Agronomy Journal. 1974;66 (2): 239-240

Iavicoli, A., Boutet, E., Buchala, A., and Métraux, J. P. 2003. Induced systemic resistance in *Arabidopsis thaliana* in response to root inoculation with *Pseudomonas fluorescens* CHA0. Mol. Plant-Microbe Interact. 16: 851-858.

Isakeit, T., "Using Topguard to control root rot: pay attention to that label," Texas Row Crops Newsletter, 2015.

Islam, T. M., Hashidoko, Y., Deora, A., Ito, T., and Tahara, S. 2005. Suppression of damping-off disease in host plants by the rhizoplane bacterium *Lysobacter* sp. strain SB-K88 is linked to plant colonization and antibiosis against soilborne peronosporomycetes. Appl. Environ. Microbiol. 71: 3786-3796.

Kang, S. E., Sumabat, L. G., Melie, T., Mangum, B., Momany, M., and Brewer M. T. 2020. Evidence for the agricultural origin of antimicrobial resistance in a fungal pathogen of humans. bioRxiv. **Article** / **Google Scholar**

Keel, C. Voisard, C., Berling, C. H., Kahir, G., and Defago, G. 1989. Iron sufficiency is a prerequisit for suppression of tobacco black root rot by *Pseudomonas fluorescnes* strain CHA0 under gnotobiotic contiditions. Phytopathology 79: 584-589.

Kerr, A. 1980. Biological control of crown gall through production of agrocin 84. Plant Dis. 64: 25-30.

Kessel, G. J. T., Köhl, J., Powell, J. A., Rabbinge, R., and Van der Werf, W. 2005. Modeling spatial characteristics in the biological control of fungi at the leaf scale: Competitive substrate colonization by *Botrytis cinerea* and the saprophytic antagonist *Ulocladium atrum*. Phytopathology 95: 439-448.

Kilic-Ekici, O., and Yuen, G. Y. 2003. Induced resistance as a mechanisms of biological control by *Lysobacter enzymogenes* strain C3. Phytopathology 93: 1103-1110.

Kim EH, Ro HM, Kim SL, Kim HS, Chung IM. Analysis of isoflavone, phenolic, soyasapogenol, and tocopherol compounds in soybean (Glycine max (L.) Merrill) germplasms of different seed weights and origins. Journal of Agricultural and Food Chemistry. 2012; 60: 6045-6055

Kiss, L. 2003. A review of fungal antagonists of powdery mildews and their potential as biocontrol agents. Pest Manag. Sci. 59: 475-483.

Kloepper, J. W, Ryu, C. M., and Zhang, S. 2004. Induce systemic resistance and promotion of plant growth by *Bacillus* spp. Phytopathology 94: 1259-1266.

Kloepper, J. W., Leong, J., Teintze, M., and Schroth, M. N. 1980. *Pseudomonas* siderophores: A mechanism explaining disease suppression in soils. Current Microbiol. 4: 317-320.

Kohle, H., K. Grossmann, T. Jabs et al., "Physiological effects of the strobilurin fungicide F 500 on plants," in *Proceedings of the Modern Fungicides and Antifungal Compounds III*, Thuringia, Germany, 2002.

Koller, W. 1992. Target Sites of Fungicide Action. CRC Press. Baca Raton, FL.

Koumoutsi, A., Chen, X. H., Henne, A., Liesegang, H., Gabriele, H., Franke, P., Vater, J., and Borris, R. 2004. Structural and functional characterization of gene clusters directing nonribosomal synthesis of bioactive lipopeptides in *Bacillus amyloliquefaciens* strain FZB42. J. Bact. 186: 1084-1096.

Lafontaine, P. J., and Benhamon, N. 1996. Chitosan treatment: an emerging strategy for enhancing resistance of greenhouse tomato to infection by *Fusarium oxysporum* f.sp. *radici-lycopersici*. Biocontrol Sci. Technol 6: 111-124.

Leclere, V., Bechet, M., Adam, A., Guez, J. S., Wathelet, B., Ongena, M., Thonart, P., Gancel, F., Chollet-Imbert, M., and Jacques, P. 2005. Mycosubtilin overproduction by *Bacillus subtilis* BBG100 enhances the organism's antagonistic and biocontrol activities. Appl. Environ. Microbiol. 71: 4577-4584.

Leeman, M., Van Pelt, J. A., Den Ouden, F. M., Heinbroek, M., and Bakker, P. A. H. M. 1995. Induction of systemic resistance by *Pseudomonas fluorescens* in radish cultivars differing in susceptibility to Fusarium wilt, using novel bioassay. Eur. J. Plant Pathol. 101: 655-664.

Leeman, M., Van Pelt, J. A., Hendrickx, M. J., Scheffer, R. J., and Bakker, P. A. H. M. 1995. Biocontrol of Fusarium wilt of radish in commercial greenhouse trials by seed treatment with *Pseudomonas fluorescens* WCS374. Phytopathol 85: 1301-1305.

Linderman, R. G. 1994. Role of AM fungi in biocontrol. Pages 1-25 in: Mycorrhizae and Plant Health. F. L. Pfleger and R. G. Linderman, eds. APS Press, St. Paul, MN.

Lindsay, W. L. 1979. Chemical Equilibria in Soils. John Wiley & Sons, Inc., New York.

Loper, J. E., and Buyer, J. S. 1991. Siderophores in microbial interactions of plant surfaces. Mol. Plant-Microbe Interact. 4: 5-13.

Lusas EW. Soybean Processing and Utilization. In: Boerma, H.R. and Specht, J.E., editors. Soybeans: Improvement, Production and Uses. Medison (USA): ASA, CSSA, SSSA, 2004. p. 949-1045

Lyr, H., Russell, P.E., and Sisler, H.D. 1996. Modern Fungicides and Antifungal Compounds. Intercept Limited. Andover, United Kingdom.

Maurhofer, M., Hase, C., Meuwly, P., Métraux, J. P., and Defago, G. 1994. Induction of systemic resistance to tobacco necrosis virus by the root-colonizing *Pseudomonas fluorescens* strain CHA0: influence of the *gac*A gene and of pyoverdine production. Phytopathology 84: 139-146.

McBride WD, Greene C. The profitability of organic soybean production. Renewable Agriculture and Food Systems. 2009, 24 (4): 276-284

McKellar, M. E., and Nelson, E. B. 2003. Compost-induced suppression of *Pythium* damping-off is mediated by fatty-acid-metabolizing seed-colonizing microbial communities. Appl. Environ. Microbiol. 69: 452-460.

McSpadden Gardener, B., and Fravel, D. 2002. Biological control of plant pathogens: Research commercialization, and application in the USA. Online. Plant Health Progress doi: 10.1094/PHP-2002-0510-01-RV.

Meziane, H., Van der Sluis, I., Van Loon, L. C., Hofte, M., and Bakker, P. A. H. M. 2005. Determinants of *Pseudomonas putida* WCS358 involved in inducing systemic resistance in plants. Mol. Plant Pathol. 6: 177-185.

Mihajlovic, M., E. Rekanovic, J. Hrustic et al., "In vitro and in vivo toxicity of several fungicides and Timorex gold biofungicide to *Pythuim aphanidermatum*," *Pesticides & Phytomedicine*, vol. 28, no. 2, pp. 117–123, 2013.

Milgroom, M. G., and Cortesi, P. 2004. Biological control of chestnut blight with hypovirulence: a critical analysis. Annu. Rev. Phytopathol. 42: 311-338.

Milgroom, M., and Cortesi, P. 2004. Biological control of chestnut blight with hypoviulence: A critical review. Annu. Rev. Phytopathol. 42: 311-338.

Morris, P. F., and Ward, E. W. R. 1992. Chemoattraction of zoospores of the plant soybean pathogen, *Phytophthora sojae*, by isoflavones. Physiol. Mol. Plant Pathol. 40: 17-22.

Morton, J. B., and Benny, G. L. 1990. Revised classification of arbuscular mycorrhizal fungi (zygomycetes): a new order glomales, two new suborders, glomineae and gigasporineae and gigasporaceae, with an amendation of glomaceae. Mycotaxon 37: 471-491.

Moyne, A. L., Shelby, R., Cleveland, T. E., and Tuzun, S. 2001. Bacillomycin D: an iturin with antifungal activity against *Aspergillus flavus*. J. Appl. Microbiol. 90: 622-629.

Mueller, D., "Evaluation of foliar fungicides for management of soybean rust," *Integrated Crop Management* IC-498, no. 3, 2007.

Mueller, D.S., Wise, K.A., Sisson, A.J., Allen, T.W., Bergstrom, G.C., Bissonnette, K.M., Bradley, C.A., Byamukama, E., Chilvers, M.I., Collins, A.A., Esker, P.D., Faske, T.R., Friskop, A.J., Hagan, A.K., Heiniger, R.W., Hollier, C.A., Isakeit, T., Jackson-Ziems, T.A., Jardine, D.J., Kelly, H.M., Kleczewski, N.M., Koehler, A.M., Koenning, S.R., Malvick, D.K., Mehl, H.L., Meyer, R.F., Paul, P.A., Peltier, A.J., Price, P.P., Robertson, A.E., Roth, G.W., Sikora, E.J., Smith, D.L., Tande, C.A., Telenko, D. E. P., Tenuta, A.U., Thiessen, L.D., and Wiebold, W.J. 2020. Corn yield loss estimates due to diseases in the United States and Ontario, Canada, from 2016 to 2019. Plant Health Progress.

Neilands, J. B. 1981. Microbial iron compounds. Annu. Rev. Biochem. 50: 715-731.

Nelson, D. W. and L. E. Sommers, "Determination of total nitrogen in plant material," *Agronomy Journal*, vol. 65, no. 1, pp. 109–112, 1973.

Nene , Y. L. and Thapliyal, P. N. 1993. Fungicides in Plant Disease Control. Science Publishers, Inc. India.

Notz, R., Maurhofer, M., Schnider-Keel, U., Duffy, B., Haas, D., and Defago, G. 2001. Biotic factors affecting expression of the 2,4-diacetylphloroglucinol biosynthesis gene phlA in *Pseudomonas fluorescens* biocontrol strain CHA0 in the rhizosphere. Phytopathology 91: 873-881.

Numberger, T., Brunner, F., Kemmerling, B., and Piater, L. 2004. Innate immunity in plants and animals: striking similarities and obvious differences. Immunological Rev. 198: 249-266.

Odum, E. P. 1953. Fundamentals of Ecology. W. B. Saunders, Philadelphia / London.

Oliver, R. P. and Hewitt, H. G. 2014. Fungicides in Crop Protection. 2nd edition. CABI Publishing, CAB International.

Ongena, M., Duby, F., Rossignol, F., Fouconnier, M. L., Dommes, J., and Thonart, P. 2004. Stimulation of the lipoxygenase pathway is associated with systemic resistance induced in bean by a nonpathogenic *Pseudomonas* strain. Mol. Plant-Microbe Interact. 17: 1009-1018.

Ordentlich, A., Elad, Y., Chet, I. 1988. The role of chitinase of *Serratia marcescens* in the biocontrol of *Sclerotium rolfsii*. Phytopathology 78: 84-88.

Page, B. G. and Thomson, W.T. 2003. The 2003 Newly Revised Insecticide, Herbicide, Fungicide Quick Guide. Kovak Books. Bakersfield, CA.

Palumbo, J. D., Yuen, G. Y., Jochum, C. C., Tatum, K., and Kobayashi, D. Y. 2005. Mutagenesis of beta-1,3-glucanase genes in Lysobacter enzymogenes strain C3 results in reduced biological control activity toward Bipolaris leaf spot of tall fescue and Pythium damping-off of sugar beet. Phytopathology 95: 701-707.

Paul, P. A., Bradley, C. A., Madden, L. V., Dalla Lana, F., Bergstrom, G. C., Dill-Macky, R., Wise, K. A., Esker, P. D., McMullen, M. P., Grybauskas, A., Kirk, W. W., Milus, E. A., and Ruden, K. 2018. Effects of pre- and post-anthesis applications of demethylation inhibitor fungicides on Fusarium head blight and deoxynivalenol in spring and winter wheat. Plant Disease 102: 2500-2510.

Paul, P. A., L. V. Madden, C. A. Bradley et al., "Meta-analysis of yield response of hybrid field corn to foliar fungicides in the U.S. corn belt," *Phytopathology*, vol. 101, no. 9, pp. 1122–1132, 2011.

Paulitz, T. C., and Belanger, R. R. 2001. Biological control in greenhouse systems. Annu. Rev. Phytopathol. 39: 103-133.

Petit, A.-N., F. Fontaine, P. Vatsa, C. Clément, and N. Vaillant-Gaveau, "Fungicide impacts on photosynthesis in crop plants," *Photosynthesis Research*, vol. 111, no. 3, pp. 315–326, 2012.

Pieterse, C. M. J., Van Wees, S. C. M., Ton, J., Van Pelt, J. A., and Van Loon, L. C. 2002. Signalling in rhizobacteria induced systemic resistance in *Arabidopsis thaliana*. Plant Biol. 4: 535-544.

Portillo, H. E., R. R. Evans, J. S. Barnes, and R. E. Gold, "F500, a new broad-spectrum fungicide for control of peanut diseases," *Phytopathology*, vol. 91, p. S202, 2001.

Press, C. M., Loper, J. E., and Kloepper, J. W. 2001. Role of iron in rhizobacteria mediated induced systemic resistance of cucumber. Phytopathology 91: 593-598.

Price III, P. P., Purvis, M. A., Cai, G., Padgett, G. B., Robertson, C. L., Schneider, R. W., and Albu, S. 2015. Fungicide resistance in *Cercospora kikuchii*, a soybean pathogen. Plant Disease 99: 1596-1603.

Raaijmakers, J. M., Vlami, M., and De Souza, J. T. 2002. Antibiotic production by bacterial biocontrol agents. Anton. van Leeuw. 81: 537-547.

Rajcan I, Hou G, Weir AD. Advances in Breeding of Seed Quality Traits in Soybean. In: Kang MS editor. Genetic and production Innovations in Field Crop Technology: New Development in Theory and Practice. Food Product Press, 2005. p. 145-174

Rebetzke GJ, Burton JW, Carter Jr EC, Wilson RF. Changes in Agronomic and Seed Characteristics with Selection for Reduced Palmitic Acid Content in Soybean. Crop Science. 1998, 38(2): 297-302

Rondon, M. N. and Lawrence, K. S. 2019. *Corynespora cassiicola* isolates from soybean in Alabama detected with the G143A mutation in the cytochrome b gene. Plant Health Progress 20: 247-249.

Ross, E. W., and Marx, D. M. 1972. Susceptibility sand pine to *Phytophthora cinnamomi.* Phytopathology 62: 1197-1200.

Ryu, C. M., Farag, M. A., Hu, C. H., Reddy, M. S., Kloepper, J.W., and Pare, P. W. 2004. Bacterial volatiles induce systemic resistance in *Arabidopsis*. Plant Physiol. 134: 1017-1026.

Sandra, A. I., Wright, C. H., Zumoff, L. S., and Steven, V. B. 2001. *Pantoea agglomerans* strain EH318 produces two antibiotics that inhibit *Erwinia amylovora* in vitro. Appl. Environ. Microbiol. 67: 282-292.

Shanahan, P., O'Sullivan, D. J., Simpson, P., Glennon, J. D., and O'Gara, F. 1992. Isolation of 2,4-Diacetylphloroglucinol from a fluorescent pseudomonad and investigation of physio-logical parameters influencing its production. Appl. Environ. Microbiol. 58: 353-358.

Shaw EJ, Rajcan I. Molecular Mapping of Soybean Seed Tocopherols in the cross OAC Bayfield x OAC Shire. Plant Breeding. 2017, 136: 83-93

Sheldrick, B. H., "Test of the LECO CHN-600 determinator for soil carbon and nitrogen analysis," *Canadian Journal of Soil Science*, vol. 66, no. 3, pp. 543–545, 1986.

Smith, I.M. 1986. Fungicides for Crop Protection: 100 Years of Progress. International Specialized Book Services. Portland, OR.

Smith, K. P., Havey, M. J., and Handelsman, J. 1993. Suppression of cottony leak of cucumber with *Bacillus cereus* strain UW85. Plant Dis. 77: 139-142.

Sneh, B., Dupler, M., Elad, Y., and Baker, R. 1984. Chlamydospore germination of *Fusarium oxysporum* f. sp. *cucumerinum* as affected by fluorescent and lytic bacteria from *Fusarium* suppressive soils. Phytopathology 74: 1115-1124.

Spolti, P., Del Ponte, E. M., Dong, Y., Cummings, J. A., and Bergstrom, G. C. 2014. Triazole sensitivity in a contemporary population of *Fusarium graminearum* from New York wheat and competitiveness of a tebuconazole-resistant isolate. Plant Disease 98: 607-613.

Stierl, R., E. J. Butterfield, H. Koehle, and G. Lorenz, "Biological characterization of the new strobilurin fungicide BAS 500 F," *Phytopathology*, vol. 90, p. S74, 2000.

Stierl, R., M. Merk, W. Schrof, and W. S. Butterfield, "Activity of the new BASF strobilurin fungicide, BAS 500 F, against *Septoria tritici* on wheat," in *Proceedings of the British Crop Protection Council Conference*, pp. 859–864, Brighton, UK, 2000.

Stockwell, V. O., Johnson, K. B., Sugar, D., and Loper, J. E. 2002. Antibiosis contributes to biological control of fire blight by *Pantoea agglomerans* strain Eh252 in orchards. Phytopathology 92: 1202-1209.

Sweeney, R. A., "Generic combustion method for determination of crude protein in feeds: collaborative study," *Journal of the Association of Official Analytical Chemists*, vol. 72, no. 5, pp. 770–774, 1989.

Swoboda, C. and P. Pedersen, "Effect of fungicide on soybean growth and yield," *Agronomy Journal*, vol. 101, no. 2, pp. 352–356, 2009.

Tari, P.H., and Anderson, A. J. 1988. Fusarium wilt suppression and agglutinability of *Pseudomonas putida*. Appl. Environ. Microbiol. 54: 2037-2041.

Telenko, D. E. P., Ravellette, J. D., and Wise, K. A. 2020. Assessing late vegetative and tasseling fungicide application timings on foliar disease and yield in Indiana corn. Plant Health Progress. 21: 224-229.

Telenko, D., Chilvers, M., Robertson, A., Tenuta, A., and Smith, D. 2021. Will a second fungicide be worth the cost for tar spot. CPN 2018.

Thomashow, L. S., and Weller, D. M. 1988. Role of a phenazine antibiotic from *Pseudomonas fluorescens* in biological control of *Gaeumannomyces graminis* var. *tritici*. J. Bact. 170: 3499-3508.

Thomashow, L. S., Bonsall, R. F., and Weller, D. M. 2002. Antibiotic production by soil and rhizosphere microbes *in situ*. Pages 638-647 in: Manual of Environmental Microbiology (2nd ed.), ASM Press, Washington DC.

Thomashow, L. S., Weller, D. M., Bonsall, R. F., and Pierson, L. S. III. 1990. Production of the antibiotic phenazine-1-carboxylic acid by fluorescent pseudomonas in the rhizosphere of wheat. Appl. Environ. Microbiol. 56: 908-912.

Thomson, S. V., and Gouk, S. C. 2003. Influence of age of apple flowers on growth of *Erwinia amylovora* and biological control agents. Plant Dis. 87: 502-509.

US Congress Office of Technology Assessment. 1995. Biologically-based technologies for pest control. OTA-ENV-636. US Government Printing Office, Washington, DC.

Vallad, G. E., and Goodman, R. M. 2004. Systemic acquired resistance and induced systemic resistance in conventional agriculture: review and interpretation. Crop Sci. 44: 1920-1934.

van Dijk, K., and Nelson, E. B. 2000. Fatty acid competition as a mechanism by which *Enterobacter cloacae* suppresses *Pythium ultimum* sporangium germination and damping-off. Appl. Environ. Microbiol. 66: 5340-5347.

Van Loon, L. C., Bakker, P. A. H. M., and Pieterse, C. M. J. 1998. Systemic resistance induced by rhizosphere bacteria. Annu. Rev. Phytopathol. 36: 453-483.

Van Peer, R., and Schippers, B. 1992. Lipopolysaccharides of plant-growth promoting *Pseudomonas* sp. strain WCS417r induce resistance in carnation to Fusarium wilt. Neth. J. Plant Pathol. 98: 129-139.

Van Wees, S. C. M., Pieterse, C. M. J., Trijssenaar, A., Van't Westende, Y., and Hartog, F. 1997. Differential induction of systemic resistance in Arabidopsis by biocontrol bacteria. Mol. Plant-Microbe Interact. 10: 716-724.

Velasco L, Fernandez-Martinez JM. Breeding Oilseed Crops for Improved Oil Quality. In: Basra, A.S. and Randhawa, L.S. editors. Quality Improvement in Field Crops. Oxford, Food Product Press, 2002. p. 309-344

Vincelli, P. and D. Hershman, "Update on fungicides for 'plant health,'" *Kentucky Pest News*, no. 1212, 2009.

Voisard, C., Keel, C., Haas, D., and Defago, G. 1989. Cyanide production by *Pseudomonas fluorescens* helps suppress black root of tobacco under gnotobiotic conditions. EMBO J. 8: 351-358.

Vollmann J, Sagara T, Pachner M, Schmidt A, Mayer HK. The polyamine spermidine as a health component in soy-foods: From soybean breeding to food production. International Legume Society and Institute of Plant Genetics, Polish Acdemy of Sciences editors. Third International Legume Society Conference Legumes for human health and planet health. 21-24 May 2019, Poznan, Poland Book of Abstracts, ISBN 978-83-950380-1-3. p. 163

Watanabe D, Lošak T, Vollmann J. From proteomics to ionomics: Soybean genetic improvement for better food safety. Genetika. 2018, 50: 333-350

Weller, D. M., and Cook, R. J. 1983. Suppression of take-all of wheat by seed treatments with fluorescent pseudomonads. Phytopathology 73: 463-469.

Weller, D. M., Raaijmakers, J., McSpadden Gardener, B., and Thomashow, L. M. 2002. Microbial populations responsible for specific soil suppressiveness to plant pathogens. Annu. Rev. Phytopathol. 40: 309-348.

Wilhite, S. E., Lumsden, R. D., and Strancy, D. C. 2001. Peptide synthetase gene in *Trichoderma virens*. Appl. Environ. Microbiol. 67: 5055-5062.

Wilson RF. Seed Composion. In: Boerma, H.R. and Specht, J.E., editors. Soybeans: Improvement, Production and Uses. Medison (USA): ASA, CSSA, SSSA, 2004. p. 621-677

Windels, C. E. and J. R. Brantner, "Early-season application of azoxystrobin to sugarbeet for control of *Rhizoctonia solani* AG 4 and AG 2-2," *Journal of Sugarbeet Research*, vol. 42, no. 1, pp. 1–16, 2005.

Wu, Y.-X. and A. Von Tiedemann, "Physiological effects of azoxystrobin and epoxiconazole on senescence and the oxidative status of wheat," *Pesticide Biochemistry and Physiology*, vol. 71, no. 1, pp. 1–10, 2001.

INDEX

www.ingramcontent.com/pod-product-compliance
Lightning Source LLC
LaVergne TN
LVHW062227200926
848484LV00006B/1491